Worlds-Antiworlds

Worlds-Antiworlds

Antimatter in Cosmology

Hannes Alfvén

The Royal Institute of Technology, Stockholm

Translated by Rudy Feichtner

W. H. FREEMAN AND COMPANY

San Francisco and London

This book was first published in Stockholm,
in 1966, under the title *Världen—spegelvärlden: Kosmologi och antimateria* by Bokförlaget Aldus/Bonniers.

Printed in the United States of America.

Library of Congress Catalog Card Number 66-27947

(C3)

ACKNOWLEDGMENT

The theory of the development of the metagalactic system described in this book originated with Doctor O. Klein, former Professor of Theoretical Physics at the University of Stockholm.

Contents

I

Cosmology and Natural Science

How did the world begin? How did it become what it is now? Questions like these have intrigued the curiosity of man since his earliest days. Prescientific civilizations gave their answers to these questions in the form of cosmological myths, which describe how the gods created the world out of nothing or brought order to a primordial chaos from which the gods themselves had originated. Such myths also tell how the world was governed and put into its present condition at the councils of all-powerful deities.

These ancient myths were more than the products of unrestrained imagination. They were often based on observations that had been made during a long period of time, and they incorporated much of what was already known about the world. As a result, cosmology became an elaborate tissue of fact and fantasy.

Of special significance were the astronomical observations. It was evident that the celestial phenomena governed the terrestrial. The sun's movements controlled the succession of day and night and the change of seasons, and the moon regulated the tides and perhaps the process of menstruation as well. But what about the other heavenly bodies? Surely, didn't these too exert their influence on man and on the course of his earthly history? It seemed extremely important to find out.

This was the beginning of astrology, which maintained that human affairs were determined by the stars. However, it also marked the beginning of astronomy, according to which the stars move in observance of certain fixed laws. As astronomy developed, its practitioners learned to predict solar eclipses and other celestial phenomena. This was a significant breakthrough, because it implied invariable conformity to law independent of arbitrary whim. Therefore, if human destiny was subject to the stars and the motion of stars was governed by natural laws, how much latitude was left to omnipotent and capricious gods?

During the ensuing thousands of years, as more and more astronomical data were gathered, an enormous amount of human ingenuity was expended in trying to reconcile fact and myth into a harmonious world picture. A few hundred years ago, however, two new factors began to assert themselves.

First, craftsmen learned to grind lenses and assemble them into telescopes, which increased immensely the human powers of observation. The advancing cleverness of the human hand made the human eye more penetrating.

Second, a number of philosophers and scholars undertook the study of apparently simple physical problems—of how pendulums swing and balls roll. What could they find in such trivia to absorb them? Why didn't they continue to plumb the fascinating mysteries of the universe? The reason is simple enough: by investigating such very ordinary phenomena they could, within one province at least, free themselves from the flimsiness of myth and build up a system of knowledge that stemmed solely from verifiable observations. No experiment was acceptable unless it was "reproducible"—that is, would produce an identical result regardless of when, where, or by whom it was made. The results of different experiments were gathered together, and from them a synthesis of all the observations was developed into a theory. However, a theory is not supposed to "explain" certain phenomena once and for all, but rather to coordinate them into a systematic body of knowledge. Further, a theory should make it possible to predict the result of new experiments in advance. Indeed, the validity of a theory can be tested only by experiments.

A scientific theory must therefore not contain any elements of metaphysics or mythology. An attempted picture of the universe should embrace a logical synthesis of the observations, with all guesswork left out. Ever since this tenet was made a golden rule of science, the drawing of sharp distinctions between fact and myth has become absolutely fundamental. One of the most important tasks of the natural scientist has been to eliminate myths and prejudices. This "watchdog" duty is no less imperative today, especially since our contemporary myths like to garb themselves in scientific dress in pretense of great respectability.

The solid structure of natural science therefore gives us a foundation on which to build in our search for answers to the questions "How did the world begin?" and "How did it develop?" By contrast with most other fields of knowledge, however, we necessarily labor under severe difficulties. We continually run up against the questions "What exists beyond that?" and "What happened before then?"—questions that lie on the very frontier of our possible knowledge. We don't know how the universe "originally" arose, and perhaps we shall never know for sure. It may be one of those questions that are essentially meaningless. This is a matter for the philosopher to determine, for it falls outside the scientist's purview.

What *is* essential to our thinking when we come to the outermost fringes of astronomy is to avoid getting ensnared in a jungle where facts and myths are intertwined, as they were in the speculations of past centuries. To formulate a general rule for such avoidance is difficult, perhaps impossible. We can nevertheless say that we are working with this objective in mind when we try to incorporate the maximum of astronomical facts into the body of laboratory physics. This means that the natural laws discovered by physicists in their laboratories should be our starting point, and that we should apply these laws to the understanding of astronomical phenomena.

But can we simply assume that the laws discovered in a small earth-bound laboratory are identical with those which steer the tremendous events in outer space? This is a reasonable objection. Every new astronomical discovery, of course, tends to make us even more crushingly aware of our own insignificance. Is it not presumptuous for us to believe that our natural laws—for all the incessant experimentation and cogitation that has gone into their making—hold true even millions and billions of light-years away? Such reasoning, however, does not help us decide whether "terrestrial" laws are applicable to "celestial" phenomena. Nor is logical analysis of much use either, be it ever so acute. We must attempt to use our "terrestrial laws," and then evaluate the result. Our natural science is essentially empirical.

Time and again, from one subdivision of astronomy to another, cosmic events have appeared to obey the same natural laws that hold on earth. The moon's motion is guided by the same forces that made the apple fall from Newton's tree. The motions of stars in our galaxy comply with the same general laws as the pendulum and gyroscope. The atoms of remote stars emit the same spectral lines as the atoms analyzed in a laboratory. These examples could be multiplied. At all events, present comparisons between astronomy and physics do not disclose the existence of any phenomenon in the cosmos that would compel us to introduce new laws of nature.

Even so, whenever a new phenomenon is encountered in astrophysics, and perhaps particularly in cosmology, many excitedly shout the equivalent of "Eureka, I've found a new law!" It is an easy temptation to succumb to: if the existing laws do not seem to fit satisfactorily, then postulate a new one to explain the phenomenon. But this way of proceeding readily leads to logical chaos. It would be like a chess game in which the players kept changing the rules. We must, obviously, always be prepared to accept new natural laws if their adoption proves necessary. It is certainly more reasonable, however, to determine how many of our cosmic observations we can array under the system of laws we have found by means of well-controlled laboratory experiments. Should it then emerge that certain cosmic facts directly conflict with the laws of terrestrial physics, the next step would be to discuss the kinds of new laws which need to be introduced. We are not in that situation at the present time.

If we start from the laws of physics, we have a firm foundation for our analysis. But having said this, we still leave room for the inevitable doubts about how these laws should be applied. The doubts will pertain not only to astrophysical and cosmological problems, but also to events on earth. No one questions the obedience of the earth's atmosphere to the laws of mechanics and atomic physics. All the same, it may be extremely difficult for us to determine how these laws operate with respect to any given situation involving atmospheric phenomena. But the laws of mechanics or atomic physics are not proved wrong because rain is falling even though the weather man predicted sunshine. His mistake merely shows that these laws are very difficult to apply in meteorology.

The same difficulty holds true for cosmology.

II

What Does the World Consist Of?

MICROSCOPIC AND MACROSCOPIC PHENOMENA

Natural science today is mainly concerned with two areas of investigation: the macroscopic (ultimately represented by astronomical phenomena) and the microscopic (ultimately represented by the elementary particles of nature); see Figure 1. Since the study of cosmological problems is our objective, we begin by reviewing in this chapter that part of our astronomical knowledge which is relevant for this purpose. However, the microscopic world provides indispensable terms of reference for understanding the macroscopic world, so in the next chapter we review some pertinent aspects of particle physics. Furthermore, the behavior of a collection of particles under conditions often found in the cosmic environment is of major importance in understanding astrophysical and cosmological phenomena, so in Chapter IV we review the appropriate aspects of plasma physics.

SOLAR SYSTEMS AND GALAXIES

Let us begin with a brief recapitulation of what we know about our astronomical surroundings. We shall have frequent occasion to specify a distance in terms of the time it takes light to travel

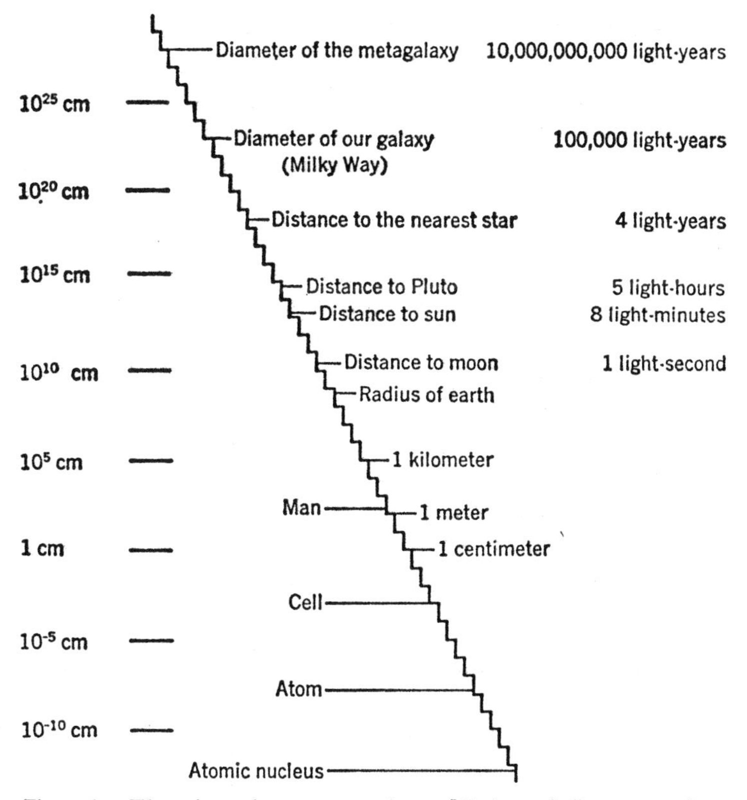

Figure 1. The staircase from atom to universe. [Each step indicates a ten times enlargement. Example: an object 5 steps higher than another is larger by $10 \times 10 \times 10 \times 10 \times 10 = 100{,}000$ times.]

a given length. The speed of light is 300,000 kilometers per second. We might then express the size of the earth in terms of the time it would take a ray of light to travel around it; this is one-seventh of a second. A ray of light can travel from earth to moon in slightly more than one second, and for the sake of convenience we speak of this distance as a "light-second." The sun is eight light-minutes from the earth. Pluto, the outermost planet, moves at a distance of five light-hours from the sun. An artificial satellite, traveling at a speed that would carry it once around the earth every 90 minutes, would take eight months to reach the sun, and 25 years to reach Pluto.

Even so, the space within our planetary system may be regarded as the astronomer's own backyard, for beyond our solar system space manifests an overwhelming scale. The sun's closest stellar neighbor, Alpha Centauri, is four light-years away; a spaceship traveling at present-day artificial satellite speeds would take more than 100,000 years to get there. The stars we see at night are seldom less than 10 light-years apart. (Many of them are binaries—that is, two stars relatively close together and revolving around each other; and many of the stars, no doubt, are the centers of planetary systems that may contain organic life.)

All the stars that we see comprise a gigantic lens-shaped system of stars known as the Milky Way or, simply, the *galaxy*. It consists of no less than 10^{11} stars, of the same average size as our sun. However, they are not evenly distributed in space but often

Table 1. Extremely Large and Extremely Small Numbers

				Prefix	
One trillion	=	1 000 000 000 000	= 10^{12}	tera	T
One billion	=	1 000 000 000	= 10^{9}	giga	G
One million	=	1 000 000	= 10^{6}	mega	M
One thousand	=	1 000	= 10^{3}	kilo	k
One thousandth	=	0.001	= 10^{-3}	milli	m
One millionth	=	0.000 001	= 10^{-6}	micro	μ
One billionth	=	0.000 000 001	= 10^{-9}	nano	n
One trillionth	=	0.000 000 000 001	= 10^{-12}	pico	p

Examples

One hundred million is a figure with 8 zeros after the one and is written as 10^8 (called "ten to the eighth").

One hundred millionth has 8 zeros before the one and is written as 10^{-8} (called "ten to the minus eight").

Five thousandth of a meter is written as $5 \cdot 10^{-3}$ meter or 5 *milli*meters.

Two billion volts is written as $2 \cdot 10^9$ volt or 2 GV or 2 BV (called "two gigavolt" or "2 BV").

Figure 2. The Andromeda nebulosa. [Photograph from the Mount Wilson and Palomar Observatories.]

Figure 3. Galaxy with spiral structure. [Photograph from the Mount Wilson and Palomar Observatories.]

gathered in *clusters*. The galactic lens has a diameter of 100,000 light-years (10^5) and a thickness of 10,000 light-years (10^4). Our solar system is far from the center, about halfway to the periphery.

The next big step toward infinity brings up the question: Is there anything beyond our galaxy? In the constellation of Andromeda, which shines high in the northern heavens during autumn, the naked eye can make out a foggy object. This is the Great Nebula in Andromeda (see Figure 2), which turns out to be a system of stars roughly similar in size and shape to our galaxy. It is about two million light-years away (2×10^6).

A vast number of other galaxies have been revealed by the giant telescopes. They vary in size and appearance, but the figures we mentioned give an approximate idea of their dimensions and distances from one another (see Figure 3). In the same manner as individual stars group together, galaxies often form clusters.

THE METAGALACTIC SYSTEM

All these galaxies form an even larger system (by now we are beginning to run out of superlatives): the *metagalactic system*. Its size is reckoned in billions of light-years, and possibly consists of 10^{10} (ten billion) galaxies. In whatever direction we look, we find no end to it. If the system is finite—which we don't know for sure—our galaxy is located far from its bounds. We might be close to the center, or we might be quite remote from it.

When we begin to study the structure of the metagalactic system, we approach the limits of our knowledge. At the same time we enter the realm of cosmology.

THE RED SHIFT

When the light emitted by a galaxy is subjected to spectroscopic analysis, we find that most of the spectral lines agree with those of ordinary stars. It follows that the stars in outlying galaxies possess the same general chemical composition as those in our galaxy. The physical conditions prevailing in most galaxies are similar to the conditions in our own, although there are many exceptions. But a remarkable fact emerges: the lines from nearby galaxies are somewhat displaced toward the red end of the spectrum. The more remote the galaxy, the more pronounced is this displacement, which we call the *red shift*.

The shift of a spectral line toward the red signifies a lessening of light frequency. The frequencies of blue and violet light are twice that of the red.

Edwin Hubble, working at the Mount Wilson Observatory, discovered that the red shift increases in proportion to the distances of galaxies from us. The spectral lines of galaxies 10^7 light-years away have their frequencies reduced by 0.1%, and those of galaxies 10^8 light-years away by 1%. This effect holds true in whatever direction the galaxies lie.

The red shift is a phenomenon of paramount importance for cosmology, and several attempts have been made to explain it. According to the simplest explanation, the red shift depends on the "Doppler effect," which was first noted in sound waves. If a car moves toward us, the sound of its horn has a higher frequency than is actually being emitted. In everyday language, we hear a higher pitch. The percentage increase in frequency is equal to the speed of the car divided by the speed of sound. As the car moves away from us, the pitch of its horn falls off. A similar analogy applies to the changing frequency of light from a source. The light loses in frequency (that is, shifts toward the red) by, say, 1% when its source recedes from us at a speed equal to 1% of the speed of light. If the red shift of the galaxies is a Doppler effect, then the galaxies are receding at a speed proportional to their distance from us. Red shifts of more than 200,000 km per second have been observed—a fantastic rate, equivalent to more than two-thirds the speed of light.

That matter can travel so fast is in itself not shocking. In the laboratory, electrons and even atomic nuclei can be accelerated to such speeds with ease, but we have no experience of massive bodies with these speeds. Present-day artificial satellites cover about 10 km in one second. The earth travels 30 km per second in its orbit around the sun. The stars near our sun revolve around the galactic center at a rate of about 300 km per second. But if the red shift is to be interpreted as a Doppler effect, the galaxies would be moving at speeds hundreds of times faster.

At this point, however, it should be asked: Is the red shift really a Doppler effect? We know of only one other phenomenon that will produce a red shift. According to the general theory of relativity, the frequency of light which traverses a gravitational field is altered. This is because the light consists of photons (light quanta), which have "effective" mass and hence weight. Energy is therefore expended to "lift up" a photon in its gravitational field, and this is taken from the photon's own energy. Since the photon's energy is proportional to the frequency, the latter must therefore diminish, which means a red shift. (The same thing could be expressed by saying that "time goes more slowly in a gravitational field," but let us not get any deeper into the theory of relativity than we have to.) Light emitted at the surface of the sun thus shifts to the red, since when photons leave the sun they have to overcome its gravitational pull. By the time the light "falls down" on earth, it has shifted toward the violet because its energy is increased by the earth's gravitational field. The effect of this field is extremely small (though in principle it has been demonstrated by the "Mössbauer effect").

Can the red shift of the galaxies be attributed to gravitation? Let us assume that the metagalactic system is a huge sphere with a radius of some billion light-years. A photon emitted by a galaxy on the surface of this sphere, and which eventually reaches our locality, is attracted by the whole mass within the sphere and must *increase* its energy. In theory, therefore, the light in a metagalactic system ought to shift toward the violet—certainly not toward the red. (As the observed red shift is of the same magnitude in all directions, masses outside the sphere cannot produce a red shift for reasons of symmetry.) Our conclusion must be that the red shift is not a gravitational effect.

The Doppler effect is thus the only *known* physical effect that can explain the observed red shift. We cannot of course rule out the possibility that some *unknown* effect is operative. As we noted by way of introduction, it cannot simply be assumed that the laws discovered by physicists in their small earthly laboratories are identical with those that govern the immense metagalactic phenomena.

The new hypotheses introduced to explain the red shift are of two types, though they hold one feature in common: they have to do with the origin of light way back in time. We know that the light from a galaxy one billion light-years away was emitted a billion years ago. But what about the characteristics of atoms at that time? Were they the same or different? We don't know. According to Milne, the ancient atoms may have emitted lower frequencies than present-day atoms. If frequencies change proportionally with time, we could explain the galactic red shift.

The properties of atoms derive from certain physical constants. To take an example, according to Bohr's atomic theory the spectral lines of hydrogen may be calculated exactly if we know the speed of light, Planck's constant, and the mass and charge of the electron and proton. If the atoms of long ago radiated in a different way than do present-day atoms, it might be that Bohr's theory would not hold true for conditions at that time. However, this is an uninviting assumption. It is much easier to postulate that one or more of the physical "constants" are not constant, but variable with time. We could assume such a change in the speed of light or in Planck's constant. If the red shift is to be explained in these terms, we can derive the laws whereby the physical constant varies with time.

No experiment has yet succeeded in demonstrating such a change. But that is not a conclusive argument against the hypothesis, since the change needed to explain the red shift is very small, and measures of extreme accuracy have been made only in the past three decades or so. If a "natural constant" changes by 10% over a billion years, the change over 100 years amounts to a very small fraction, and enormous precision would be required to prove it. We first have to improve the accuracy of our measurements to one part in several hundred millions, and then wait a hundred years before we can prove or disprove the hypothesis.

The hypothesis on variation of physical constants enables us to draw conclusions about other phenomena. None of them, however, is definitely proved; by the same token, none can be clearly refuted.

Milne's assumption that physical constants vary is in the nature of an ad hoc assumption. It explains what it is introduced to explain—in this case the red shift—but nothing else.

Another attempt to explain the red shift proposes that atoms have always emitted the same spectral lines; it is the light that changes color during its long journey. In other words, light "gets tired" after hurtling through space for millions of years, and the fatigue manifests itself as a red shift. Light, we know, consists of a stream of photons having energy which is proportional to the frequency. A violet photon has more energy than a red photon. The loss of energy by the photons during their travels would then account for the red shift.

At first glance the assumption is fairly plausible. But when we take a closer look, serious obstacles intervene. The red shift cannot come from interaction between the photon and the matter existing in outer space, because the density of this matter is very small. All the matter through which light passes over a billion years could be comfortably fitted into less than an inch of our own atmosphere, and light rays can travel through air for miles without showing the slightest signs of "fatigue." The red shift must then derive from interaction with "empty space" of an entirely new character. Since the red shift signifies that photons lose energy, we may wonder where this energy goes. Are we being asked to renounce the conservation of energy, or shall we assume that a photon loses energy by virtue of some unknown process? Both alternatives would require a radical revision of fundamental physical concepts.

Our two attempts to explain the red shift in terms other than the Doppler effect mean that we must introduce new natural laws having the character of ad hoc assumptions. These laws have no other phenomena to support them, but neither can they be disproved. Everyone is free to accept or reject them as he pleases. We have, however, set out to inquire how far we can go without having to postulate new natural laws for which there is no incontestable evidence. We want to find out how many cosmological phenomena lend themselves to explanation in terms of the laws we have discovered by theoretical analyses of observations made in the laboratory. That leaves us with only *one* alternative: to accept the red shift as a Doppler effect. We conclude that *the galaxies are receding at a speed proportional to their distance from us.*

METAGALACTIC EXPANSION AND THE BIG-BANG THEORY

In accord with our last conclusion, it might be inferred that our galaxy occupies a special position in the universe: namely, the center. But this is not necessarily correct, as we shall soon see.

Let us consider an explosion where the fragments are hurled out in all directions and with different speeds. If each fragment maintains its own speed for all time after the explosion, its distance from the center of the explosion at any given time is directly proportional to its speed. But, as may be proved by a simple geometrical theorem, the same law applies to *any* of the fragments. An observer on one of them will notice that the others are receding at speeds proportional to their distance from him.

If we thus see that the galaxies are receding at speeds proportional to their distance from us, we may interpret this to mean that we arc somewhere not too near the edge of a collection of galaxies which were once hurled out in all directions from a big explosion. The Belgian mathematician and physicist Georges Lemaître assumed that once upon a time, billions of years ago, all matter was compressed in a single clump, an "atome primitif," which exploded. Though containing all the matter of the universe, this clump (or primordial atom) he assumed to be no larger than the distance from earth to sun, or 1/70,000 of a light-year. The matter ejected by the explosion later condensed into galaxies, and within these the stars were gradually formed (see Figure 4). The theory has been developed and modified by several astronomers, among them George Gamow, known for his often authoritative and always very entertaining expositions of popular science. Gamow introduced a number of suggestive terms, among them "ylem" for the dense primeval clump of matter and "big bang" for the explosion itself. This theory presupposes that what we call the "metagalactic system"—all the galaxies we have observed (or hope to observe)—constitutes the whole "universe." Beyond the metagalactic system there is, literally, nothing.

The time when the big bang took place can be determined

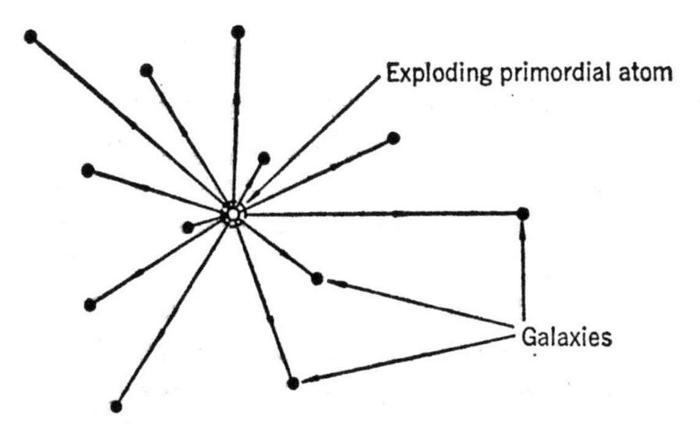

Figure 4a. The big-bang theory of the creation of the universe. According to this theory, the outward motion of the galaxies was caused by an exploding "atome primitif," which ejected them in all directions.

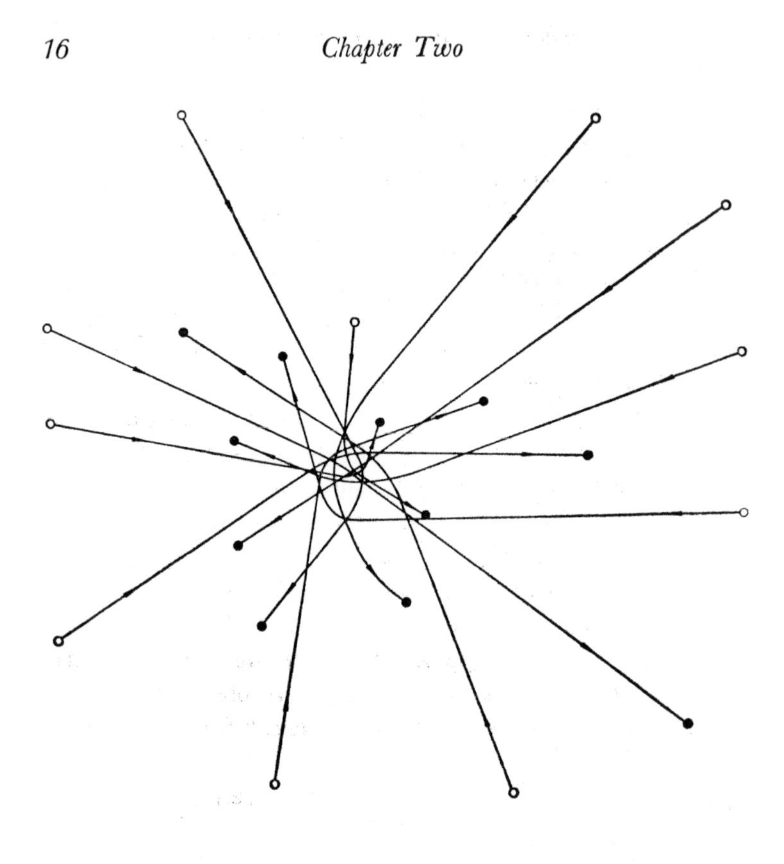

Figure 4b. Alternative to the big-bang theory: we can equally well assume that the origin is a very large cloud, whose parts have fallen in toward the center but not collided with each other. The parts of the cloud move in the orbits of the figure.

if we know the relation between the distance and speed of the galaxies, assuming no change in the speed. The date of the big bang then works out to about 10 billion years ago. But since the galaxies are mutually attracted by general gravitation, the outward speed of the exploding cloud has tended to slow down.

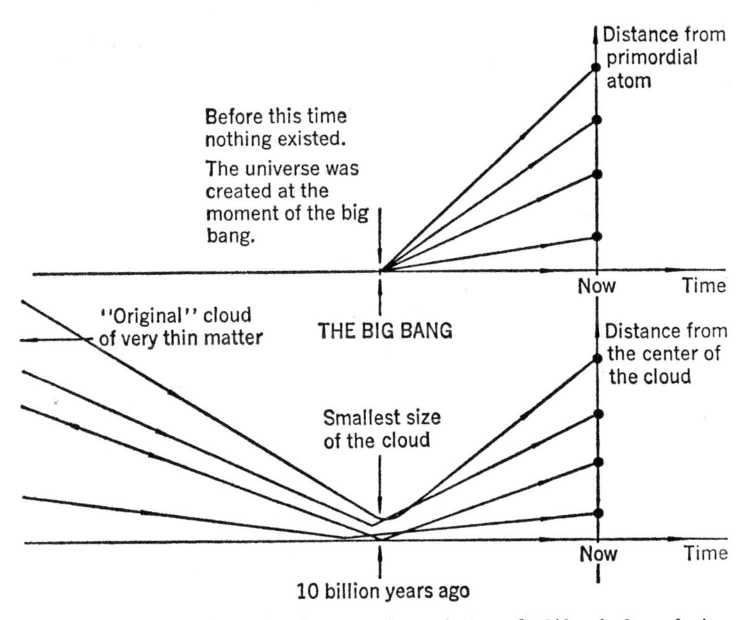

Figure 4c. Two alternative interpretations of the red shift of the galaxies. (Basic assumption: no antimatter.) Above: The galaxies were ejected when the primordial atom exploded. This interpretation presumes that the universe was created as a giant bomb 10 billion years ago. (Compare Figure 4a.) Below: The primordial state is assumed to be a very large, very diluted cloud which by gravitation contracts towards its center. Its different parts pass each other millions of light-years apart, after which they rush outward again. The minimum size was reached 10 billion years ago. (Compare Figure 4b.)

When we allow for this in our calculations, the big bang may be held to have occurred somewhat more recently.

The big-bang theory thus postulates a highly dramatic event at a fixed point of time, and it becomes natural to reckon the "age of the universe" from this point. Proponents of the theory turn extremely vague when asked about what happened before the big bang. Sometimes they suggest that a prior universe existed and may have been very much like the one we have now. It then contracted to form an ylem. But just as often we are led to believe that nothing prior existed. At a given moment, we are told, the world was created as an ylem. Gamow has even amused himself with a *minute-by-minute* account of what happened after the moment of creation!

There seems to be little justification for this vagueness about the pre-explosion state. As we shall see presently, it is scarcely probable that such a great amount of matter occupying such a small region of space as ylem can be produced from a state similar to the present one. The big bang presupposes an act of creation at a specific moment. Those who discern here the intervention of some divine power may find the theory attractive. In the beginning, nothing; out of nothing came the primordial atom, outside of which existed nothing. Certain natural laws were inherent in the act of creation; these were prescribed once and for all, and the development of the universe has since been governed by them and by nothing else.

As already pointed out, the purpose of this book is to find out how far we can come in understanding the world's structure and development with reference to the natural laws we have found in the laboratories. The belief that some supernatural force is perpetually intervening in the operation of the universe is completely foreign to this line of thinking. But even though we require that the scientific laws apply without limit to that part of the universe and during that period of time within our powers of observation, science is still hard put to answer the constantly escalating question "What lies beyond?" and "What happened before then?" It is therefore difficult to find logically tight arguments against the "great compromise," which may be formulated in this way: science took over from the moment of creation, but the primordial atom was created by supernatural intervention; and if the natural laws were also established at that time, the development of the universe was also predetermined at the moment of creation.

However, there are several alternatives to a cosmology of this type.

ALTERNATIVES TO THE BIG-BANG THEORY

We have seen that the motions of galaxies inferred from the red shift *can* be obtained from one big bang. Alternative explanations may be advanced, however. Perhaps the most interesting of them is associated with the existence of antimatter, which will be discussed in greater detail in Chapter VI. But even in a world without antimatter, an extrapolation of present galactic motions does not necessarily lead back to the occurrence of a big bang.

To simplify our discussion, let us first assume that we find ourselves at the center of the metagalactic system. Next, let us assume that the galaxies have always been receding from us at speeds equivalent to the measured red shifts. When we divide the distances of the different galaxies by their speeds, we arrive at a mean value of 10 billion years as that time in the past when all of them were located at the center. But we do not get exactly this value for every galaxy in our calculations. For one galaxy the value may be 3% higher (10.3 billion years); for another, 3% lower (9.7 billion years). But since any measurement of the red shift and of galactic distances is difficult by the very nature of things, we cannot be quite sure of these values. There may be a 5% margin of error either way. Accordingly, galaxy 1 may have been at the center anywhere from 9.8 to 10.8 billion years ago, whereas the time span for galaxy 2 would range from 9.2 to 10.2 billion years. Thus both galaxies *might* have been at the metagalactic center 10 billion years ago and *might* have stemmed from a big bang at that time. But this is by no means a *necessary* conclusion. Our observations could just as well be interpreted to mean that galaxy 1 was at the center 10.5 billion years ago and galaxy 2 was at the center 9.5 billion years ago. It follows that they need never have been near one another and need not have emanated from an exploding ylem.

We have so far assumed that the galaxies are receding from us at a speed equivalent to their red shift, and that they have been consistently moving in the same outward direction. But they may very readily be moving sideways as well. If so, they would be shifting their positions in the heavens; but since we have not been observing galaxies for much more than forty years, it would not be possible to detect such lateral motion unless it is of enormous speed. In short, the galaxies need not be moving directly away from us, but may just as well be moving at an angle to the line of sight. Even if the angle does not exceed 20° or 30°, the galaxies may have passed the metagalactic center at a very great distance from it.

In our discussion we have assumed that our galaxy occupies the center of the metagalaxy. This assumption simplifies things for us but is not essential to the result. Roughly the same conclusions may be drawn by putting our galaxy far from the center.

Thus observations of galactic motions do not necessarily lead back to a big bang. An assumed primordial atom should give a red shift according to Hubble's law, but the reverse does not hold: we cannot deduce a big bang from Hubble's law. It is just as reasonable to assume that the galaxies have passed one another at fairly great distances (perhaps averaging one-tenth or so of their present distances from the metagalactic center). The metagalaxy was smallest about 10 billion years ago, but its volume was perhaps extended over 1 billion light-years—in any case, not necessarily in a clump of 1/70,000 light-year in which the primordial atom was compacted. When we go back a further 10 billion years, the metagalaxy should have had about the same dimensions as it has now. The galaxies were then moving *toward the center* with roughly the same speeds with which they are now moving outwards. When we go back even further in time, the galaxies were dispersed over a vastly greater volume. (By now, however, we are in a period so remote that galaxies may not yet have been formed.)

The "primordial state" thus deduced stands completely opposed to the big-bang theory's extremely concentrated primordial atom. We are led instead to assume that "originally" the matter which now shapes our metagalaxy was rather uniformly distributed over a very large volume, perhaps thousands or millions or even billions of times the present volume. It is easy to imagine the primary existence of an immense mass of gas, which began to contract under the influence of its own gravitation (that is, the mutual attraction exerted by different parts of the gas cloud). As the contraction proceeded, local condensations of the gas were formed, giving rise in turn to the galaxies. Their formation, too, was the result of condensation.

If we choose to construct a mathematically simple and idealized model, we can visualize a very large sphere in which gas is uniformly distributed. Under the influence of gravitation, every part of the sphere begins to collapse toward the center. If nothing happens to disturb the collapse, according to the laws of mechanics all the parts will converge on the center at exactly the same time. An extremely dense concentration of gas would ensue. Perhaps even an ylem might be formed.

Such a model, however, would take us far from what is physically reasonable to assume. The world we live in is not mathematically perfect, and if we try to describe it with mathematical models we must always make sure that they don't stray too far from reality. In the present case it is not realistic to assume that the distribution of mass within the contracting volume is exactly homogeneous or that the volume forms a perfect sphere. Nor may all parts of the sphere be assumed to move straight toward the center without hindrance; it is more likely that irregular turbulent motions will assert themselves. As a result, the metagalaxy cannot contract into a volume as small as the big-bang theory requires. It need not even be so small that relativity effects become decisive. Once the metagalaxy reaches its narrowest dimension, which could be of the order of one billion light-years, it will begin to expand anew.

An illustration à la Gamow will show how a process of contraction as described can hardly be expected to produce an ylem. Let us picture a large number of marksmen standing in a circle and equidistant from a certain fly, all under orders to shoot at the fly when given the signal. If their bursts of fire converge with perfect precision and timing on the target, all the bullets, obviously, would meet to form one big cannonball. (We can forget about the fly!) But only the world of mathematical perfection can fabricate such cannonballs. In the real world the bullets will pass one another without colliding.

Returning to the metagalaxy, it may well be that condensation into galaxies starts to occur at an early stage. If much of the metagalactic matter is thus concentrated during the contraction phase, the galaxies may move toward the center, pass through, and then emerge along paths resembling the hyperbolas of geometry. There is some risk of collision with other galaxies, especially when the metagalaxy approaches its minimum extent, but most of them are likely to pass near the center and out again without bumping into one another. In this way the inward speeds of the galaxies, as produced by gravitation, will transform into the present outward motions as a simple result of the laws of mechanics. This we have already discussed (see Figure 4).

The theory of relativity plays a big role in all questions relating to development of the metagalaxy, but we have ignored the relativistic effects so as not to make the presentation too complicated. Besides, in its present state of development, the theory of relativity is on the whole confined to extremely simplified models, which presumably bear little relation to physically plausible states of interest in this connection. A discussion of the importance of the theory is given in Chapter VII.

SUMMARY

We began by analyzing the red shift of the galaxies and interpreted it as a Doppler effect. With the knowledge thus gained about present galactic motion, we proceeded to investigate how the motion may have arisen. The explanations turned out to fall into two main classes. According to one class, the galaxies are fragments hurled out from an explosion. This assumes that the universe first existed as a gigantic bomb, which exploded. The universe could not possibly have attained such a contracted state by virtue of previous contraction. According to the second class of explanations, a primeval state of extremely tenuous matter is postulated. Under the influence of gravity, this giant cloud of gas begins to contract. When it reaches a diameter of perhaps one billion light-years, it expands once again and the galaxies move outward as they are now doing.

The whole of our analysis presupposes that the universe contains matter of the kind normally found on earth. However, the latest findings of particle physics suggest the existence of a different type of matter, called *antimatter*. These findings may prove to have profound consequences for our image of the universe. In the following chapter, therefore, we shall consider the charac-

Table 2. Forces in Physics

1. *Nuclear forces* keep the nucleons (protons and neutrons) together in the atomic nucleus. They are the dominating forces in the nucleus, but of no importance at large distances from it.

2a. *Electric forces.* A positive charge and a negative charge attract each other, but similar charges repel. Electric forces keep the atoms together ("bind" the electrons to the nucleus). They are of a certain importance in the nucleus. At large distances the electric forces are usually not so important because of a screening effect. For example, a positive charge attracts negative charges to its neighborhood so that they screen off the field from the positive charge.

2b. *Magnetic forces* are closely related to the electric forces. Because they cannot be screened very easily, they are efficient at larger distances than the electric forces. Example: the earth's magnetic field.

3. *Gravitation* is much weaker than electric forces and therefore of no importance in the atom. As the gravitation cannot be screened, it is the dominating force at large distances. The orbits of the planets and the motions of stars and galaxies are ruled by gravitation.

teristics of matter and antimatter, after which we return to the cosmological problems.

The development of the metagalactic system is a result of the action of different forces. Table 2 gives a summary of the most important forces in physics.

III

Matter and Antimatter

THE STRUCTURE OF MATTER

The atoms of ordinary matter each comprise a positively charged, heavy nucleus, normally surrounded by one or more negatively charged, light electrons (see Figure 5). The hydrogen atom, simplest of all the atoms, contains but a single proton as its nucleus. We often use the proton's mass ($1.67 \cdot 10^{-24}$ g) as a unit and say that the nucleus of hydrogen has a mass number of 1. The proton's electric charge likewise serves as a unit, which means that the hydrogen nucleus has a charge of +1. A neutral (normal) hydrogen atom contains an electron very close to the nucleus, at a distance of about 0.5 Ångström units (1 Å = 10^{-8} cm). The electron has a charge of −1 and a mass of 1/1840 (with the proton mass as the unit).

Next simplest of the atoms is that of heavy hydrogen—also called deuterium. Its nucleus has a charge of +1 and a mass of 2, and is built up of one proton (charge +1 and mass 1) and one neutron (charge 0 and mass 1), which is an uncharged particle of about the same weight as the proton. The nucleus of heavy hydrogen is normally surrounded by one electron (charge −1), in order that the net charge of the atom be zero. The nucleus of helium consists of two protons and two neutrons,

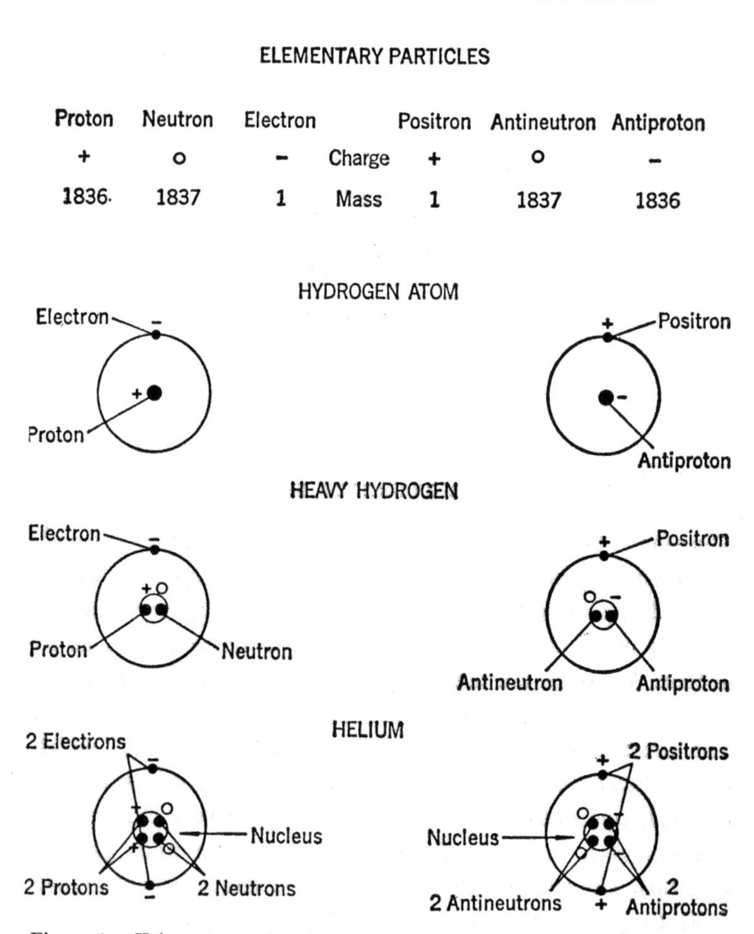

Figure 5. Koinomatter and antimatter.

giving us a charge of +2 and a mass of 4; it is surrounded by two electrons. Among the heavier elements, the nucleus of iron has a charge of +26 and a mass of 56, and is normally surrounded by 26 electrons. For uranium, the heaviest of atoms in nature, the charge is +92 and the mass 238. Even heavier atoms can be produced in particle accelerators.

ARE THE ELEMENTARY PARTICLES SYMMETRICAL?

With the development of modern atomic theory, dating from the first decades of the 20th century, there began speculation as to the import of the new conception of matter for our world picture. Basically, it was found, matter was made up of two building blocks, or elementary particles: the *proton* and the *electron*, the former being heavy and positively charged, the latter being light and negatively charged. To these was added the *neutron*, an uncharged variant of the proton. But why should the heavy particle be the positively charged one, and the light one the negatively charged? Although there is a perfect symmetry between positive and negative charge, why should there be only light particles with negative charge and only heavy particles with positive charge? This spoiled the symmetry of the world.

At this point it may be objected that there is no reason why such symmetry should be demanded of the elementary particles. To some extent, of course, the demand is made on esthetic grounds: the world would be more "beautiful"—as beauty is comprehended in mathematics or philosophy—if its basic building blocks were symmetrical in terms of charge and mass. But since the findings of early atomic research demonstrated that the building blocks were asymmetric, the experimental facts militated against the esthetic insistence on symmetry. Therefore there was nothing to do but forget about symmetry altogether, unless . . . new experimental facts could be found to prove the symmetrists right.

As it turned out, this was exactly what happened.

DISCOVERY OF THE POSITRON

As soon as the electron was discovered, speculation naturally arose about the possible existence of a positively charged electron—that is, a particle having the electron's mass (only 1/1840 that of the proton) but with a positive charge. The hunt for such a particle took on new urgency when Dirac advanced his theory of the electron. This theory, which gave a very satisfactory description of the electron's properties, required the existence of an "antielectron," a positively charged electron. A particle meeting the description was experimentally discovered in 1932, and was called the *positron*. It has the same mass (and spin) as the electron, but its charge is $+1$ instead of -1.

Positrons may be produced by gamma radiation of extremely short wavelength, which under certain conditions can "materialize" one positron and one electron. In order for this "birth of twins" to occur, the photons of which the gamma radiation consists must have energy equal to the sum of the masses of the electron and positron. According to the theory of relativity, mass and energy are equivalent. Every mass corresponds to a certain quantity of energy. Production of an electron-positron pair therefore requires a quantity of energy corresponding to at least twice the electron mass (which means about 10^6 electron volts). Since the photons are not electrically charged, the total electric charge remains unchanged (zero) if a negatively charged electron and a positively charged positron are created. The birth of an electron-positron pair is governed, as expected, by the law of conservation of the mass (equal to the energy) and the law of conservation of the electric charge.

The reverse of the birth process is also known. If an electron and a positron collide, they may "annihilate" one another. The electron's negative charge thereby neutralizes the positron's positive charge, and their total energy reappears in the form of gamma rays. Total energy is understood here to mean the energy equal to the sum of the masses of the electron and positron plus their kinetic energy.

A positron that moves in a vacuum without colliding with other particles has eternal life; the same applies to the electron. But a positron that moves through matter, whether it be a gas or solid, has a very short life. Matter contains a large quantity of electrons, and any positron that happens to collide with one of them is suddenly annihilated.

DISCOVERY OF THE ANTIPROTON

Belief in the symmetry of elementary particles was greatly strengthened by the discovery of the positron. It was accordingly expected that the proton should be complemented by an *antiproton*, a particle of the same mass but with a negative charge. Since the proton mass is 1,840 times as large as the electron mass, it takes 1,840 times more energy to produce a proton-antiproton pair than an electron-positron pair. Particle energies on the order of billions of electron volts are therefore required to generate antiprotons. To be sure, energies of this magnitude exist in cosmic rays, and many scientists claim to have found antiprotons in these rays. (And they probably did, too!) But the proof they adduced was not quite convincing. It was not until the modern giant particle accelerators went into operation that sufficiently effective sources of billion-volt energy became available to produce antiprotons.

The first clear-cut evidence for existence of the antiproton came in 1955, when the Berkeley accelerator was started. A sheet of copper was bombarded with a proton beam energized at 6.2 billion electron volts. The result: proton-antiproton pairs analogous to the formation of electron-positron pairs (similar pairs of protons and antiprotons may be generated by all kinds of high-energy radiation). The antiproton turned out to have exactly the same mass (so far as could be measured) as the proton, but its charge was negative. Like the proton, it lives forever in an absolute vacuum. But in the presence of matter it soon collides with a proton (present, of course, in every atomic nucleus) and is annihilated. This process, however, is more complicated than the electron-positron annihilation, and we shall describe it later. Just as the proton has a neutral variant, the neutron, so does the antiproton have a neutral variant, the *antineutron.* It has the same properties as the neutron, but a neutron and antineutron annihilate one another.

When the antiproton was discovered, it aroused not so much surprise as satisfaction. The symmetry of elementary particles was now an experimental fact. Both the electron and proton had antiparticles, twin brothers in every detail except for the electric charge. Nature, in short, did not distinguish between positive and negative electricity!

A number of new elementary particles were found even before the antiproton came to light, and more were due to come. The first of these were christened *mesons* and *hyperons;* like the electrons-positrons and the protons-antiprotons, they always appear in pairs. Perfect symmetry between particles and antiparticles is therefore a fundamental law of particle physics. All these particles, however, are extremely short-lived. Once created, they automatically disintegrate within a very small fraction of a second. By their very nature they cannot be a major constituent of matter, but they do figure significantly in antiproton-proton annihilation. When this happens, a number of mesons emerge. The mesons then disintegrate spontaneously within a very short time, giving rise to other types of mesons, which in turn disintegrate into electrons and positrons. These processes also cause the emission of *photons* (gamma quanta) and *neutrinos* (uncharged particles with very small mass). The usual end result of a proton-antiproton annihilation (after a fraction of one second) is one or two electrons, the same number of positrons, and gamma and neutrino radiation. If the electrons and positrons later collide, there is another annihilation (as already described), which results in further gamma radiation.

Summing up all these processes, we find that all the energy equivalent to the combined mass of the proton and antiproton is converted to radiation. Two heavy material particles have been annihilated and their mass energy has left the scene.

While on the subject of elementary particles, we should mention that the photons and neutrinos are stable particles, emanating from atoms or nuclei or from colliding elementary particles. They move at the speed of light and do not form a part of ordinary matter.

THE STRUCTURE OF ANTIMATTER

We have seen that an electron and positron annihilate one another upon contact. The same applies to a proton and antiproton, though the process of annihilation is more complicated. But what happens when an antiproton and positron are brought together? On grounds of symmetry the answer is plain and simple: the same thing that happens when a proton and electron are brought together. They form an atom.

As we noted, an ordinary atom of hydrogen has a nucleus of one proton which, being positively charged, attracts a negative electron. To borrow the terminology of Bohr's atomic theory, the negative electron may revolve around the nucleus in various different orbits. Wave mechanics prefers to say that the electron may find itself in different states of oscillation near the nucleus. When the electron shifts from one orbit to another (or from one oscillating state to another), one photon of light is emitted. By studying the spectral lines of the light, we can determine the atom's oscillating state and identify the atom.

If we replace the proton in a hydrogen atom with an antiproton, its negative charge will repel electrons but attract positrons. A positron is attracted by the antiproton with exactly the same force that acts between an electron and proton. And since the positron possesses the other properties of an electron as well, it will revolve around the antiproton in a similar manner. It may thus move in the very same Bohr orbits (or oscillate in the same states) as an electron in an ordinary hydrogen atom. A shift from one oscillating state to another emits the very same spectral line as that of an ordinary hydrogen atom. Hence the antiproton and positron form what we might call the antiatom of a new element—call it "antihydrogen"—having the same properties as ordinary hydrogen.

Two ordinary atoms of hydrogen tend to attract one another to form a molecule, consisting of two protons and two electrons. Ordinary hydrogen gas contains a large number of hydrogen molecules and, as the gas is cooled, it condenses at −252°C into liquid hydrogen. When we consider the perfect symmetry of protons-antiprotons and electrons-positrons, we feel confident that two antiprotons and two positrons can analogously form an antihydrogen molecule. If we could produce such molecules in sufficient quantity, the result would be an antihydrogen gas, which would condense into liquid antihydrogen at −252°C.

Just as a proton and neutron may combine into a deuteron, the nucleus of deuterium, an antiproton and antineutron must possess the same ability to combine into an "antideuteron," forming the nucleus of a heavy antihydrogen atom. By the same token we could expect, say, eight antiprotons and eight antineutrons to form an antioxygen nucleus, which if surrounded by eight positrons forms an antioxygen atom. Further, we could let a large number of antioxygen atoms combine with twice the number of antihydrogen atoms, giving us antiwater—a liquid which in its own right has the very same properties as water: it freezes at 0°C and boils at 100°C, turns into the most beautiful snowflakes at low temperatures, and so on. The only difference between water and antiwater is that a mixture of the two would generate tremendous energy, perhaps even an explosion. The antiprotons, antineutrons, and positrons in the antiwater would annihilate the water's protons, neutrons, and positrons. Equal quantities of water and antiwater would cancel out one another and transform into radiation.

Antihydrogen, antioxygen, and anticarbon may combine into complex organic compounds. Together with antinitrogen and several other elements they can form the whole range of chemical elements which are the bearers of organic life.

It should thus be possible to build up antimatter from antiprotons and positrons quite symmetrically with ordinary matter, the only difference being that the atomic nuclei would be negatively charged and surrounded by positrons. Should an object of antimatter be brought into contact with ordinary matter, it would act like a bomb. By being annihilated and at the same time annihilating an equal amount of matter in our world, it would release many hundred times as much energy as a hydrogen bomb of the same weight. The result would be just as violent if an object of ordinary matter were introduced into the world of antimatter.

We know in theory that it is quite possible to build up a world of antimatter in this way. Enough is known about the properties of elementary particles to enable us to say this with certainty. However, experimental techniques are not yet developed to the point where we can mass-produce antimatter. Nevertheless, some of the conclusions drawn here have been confirmed. For instance, M. Goldhaber has succeeded in producing an antideuteron, the atomic nucleus of heavy antihydrogen. But as yet no one has managed to produce a complete antiatom (with a positron rotating around an antiproton or a heavier antinucleus). Technologically, this is a formidable task; scientifically, it is of lesser interest. We can already calculate the properties of antiatoms because we are familiar with the antiprotons and positrons. We know that, if provided with sufficient laboratory resources, we could build up antimatter possessing all the properties of ordinary matter. One of the difficulties to overcome is to isolate antimatter from ordinary matter in order to prevent annihilation.

The foregoing may sound like so much science fiction. Nothing of the kind. We have been talking about experiments which improved resources will make feasible and whose results we can already predict with certainty.

But having embarked on the course of hypothetical experiments, let us continue. Suppose that we had resources enough to collect somewhere in outer space 10^{57} atoms of antihydrogen, partly mixed with miscellaneous antielements. This mass of gas would then contract under the influence of gravitation to form a star ("antistar") very much like our sun. At its center, thermonuclear processes would be generating energy and, if its chemical composition were analogous to the sun's, it would radiate light that had the same spectral lines as the sun's light.

Before we proceed, however, a pause is in order to discuss terminology. Strictly speaking, the term "antimatter" is a misnomer, since antimatter is every bit as much matter as its "ordinary" counterpart. Besides, it may well occur quite as often in nature as "ordinary" matter. Yet antimatter has become the standard term, and we have to accept it. What we can do, however, is to coin a new word for "ordinary" matter: we shall call it *koinomatter*, after the Greek word "koinos," meaning common or well known.

HOW CAN WE DISTINGUISH BETWEEN KOINOMATTER AND ANTIMATTER?

We have already mentioned the antistar, built up of antimatter. An interesting question is how—given the great distances we have to work with—we may distinguish an antistar from a koinostar, consisting of koinomatter (or ordinary matter). The only difference between koinomatter and antimatter, of course, is that the electric charges are reversed, and hence they react differently to electric and magnetic forces.

If an atom which emits light is subject to an electric field, its spectral lines are displaced (the "Stark effect"). However, in the electric fields that concern us in cosmic physics, it is never possible to induce a demonstrable Stark effect. (Formally, the pressure broadening of spectral lines is a Stark effect, but this is irrelevant for our discourse.)

The situation is different with magnetic fields. Within such a field both electrons and positrons move in circles, but they rotate in different directions. The spectral lines of a light-emitting atom situated in a magnetic field are displaced (the "Zeeman effect"), but this is different for electrons and positrons because the field makes them move in opposite directions. Assuming that we *know the direction* of a magnetic field, if we place a radiant gas in it we may determine whether the gas consists of koinomatter or antimatter by studying the Zeeman effect in the emitted light. But if we reverse the magnetic field we also reverse the direction of rotation for the electrons and positrons. In the reversed field the antimatter produces the same Zeeman effect as koinomatter did in the first field and vice versa. If we exchange the koinomatter for antimatter and at the same time turn the magnetic field, no change in the Zeeman effect results. Magnetized koinomatter can thus not be distinguished from antimatter having opposite magnetization.

Many stars are known to be magnetic because they show the Zeeman effect. We may take a certain star and assume that the direction of its Zeeman effect is such that we may infer its magnetic *north pole* to be turned toward us—provided the star consists of koinomatter. But if the star consisted of antimatter instead and was magnetized so as to turn its *south pole* toward us, the Zeeman effect would be identical. Since we have no independent method for determining whether the star is turning its north or south pole toward us, its interpretation as koinomatter or antimatter will be equally permissible.

Let us summarize the discussion so far: with two stars in outer space, the one consisting of koinomatter and the other of antimatter, we have *no way of distinguishing them on the basis of the radiation they emit.* This holds true for emitted light waves, radio waves, and gamma rays.

If the space between the stars were completely empty, observations of their light could not tell us whether the stars consisted of koinomatter or antimatter. Indeed, many of the stars, perhaps half, might be formed of antimatter and the remainder of koinomatter. But we would still be unable to distinguish between antistars and koinostars.

However, it so happens that the space between the stars is not empty. It is filled with a thin magnetized plasma, which maintains a reciprocal relationship with the stars: plasma from space may fall in toward a star, or stars may eject plasma into surrounding space. If antistars existed, they would eject antimatter, and space would be the scene of interaction between koinomatter and antimatter. What phenomenon would that produce? Could we draw conclusions as to the existence of antimatter from a study of such phenomena? To answer these questions we must turn to plasma physics, which we do in Chapter IV.

First, however, a few other questions that deserve our attention.

IS THERE ANTIMATTER IN THE SOLAR SYSTEM?

By now we have sown the seeds of doubt as to whether remote galaxies, or even relatively close stars, consist of koinomatter or antimatter. But how much of the world do we know so well that we can say, *with certainty*, that it consists of koinomatter and not antimatter?

The *earth*, by definition, consists of koinomatter. The only antimatter we have is that produced by the giant accelerators, and then only in very small amounts. In addition, a few extremely short-lived antiparticles are produced when cosmic rays hit the earth's atmosphere.

The *moon* consists of koinomatter. If it were otherwise, the moon-probing rockets would have exploded violently upon impact, and the explosion would have been readily observable on earth. No Luna or Surveyor could have been standing on its surface and sending us photographs.

The *sun* consists of koinomatter. We can be certain of this because the sun ejects plasma which, upon reaching earth, causes the aurora borealis and other phenomena. If the sun consisted of antimatter it would emit antimatter plasma, or antiplasma, and the aurorae would then glow at a thousand times their present luminosity.

The solar plasma also reaches Mercury, Venus, and Mars. We should expect to see easily observable annihilation phenomena if any of these planets consisted of antimatter. As for the outer planets, our conclusion does not hold with the same force because we are not sure whether the sun's plasma eruptions reach that far out. Even so, we can be pretty sure that koinomatter goes into the making of these planets, too. They were probably created at the same time as the remaining solar system from the same primeval blob of gas.

The whole solar system thus probably consists of koinomatter. On one point, however, we should perhaps make a reservation.

METEORS OF ANTIMATTER?

We cannot with certainty exclude that there may be small bodies of antimatter within the confines of our solar system. If so, they would come from remote regions where antimatter is present.

A star that, like our sun, has planets revolving around it, may eject solid bodies into the surrounding space. The space between the planets of our solar system contains many small bodies, from asteroids and comets down to the finest dust. All of them normally move in ellipses (Kepler orbits) around the sun. Their motions, however, are disturbed by the planets. If one of them came close to Jupiter, say, it might be perturbed out of its original orbit with sufficient additional velocity to leave the solar system. It would then have some chance to reach the vicinity of other stars.

Let us assume that an antistar with the activity described is only a few light-years distant. Conceivably, it could hurl our way bodies of antimatter which would enter our solar system.

Such a body would be observable as a meteor upon contact with the earth's atmosphere. However, the existence of antimatter in a meteor has never been demonstrated with certainty. Libby has suggested that the meteor which fell on Siberia in 1908 consisted of antimatter. This possibility has not been disproved, but neither has it been proved—as yet.

IV
Plasma Physics

THE PROPERTIES OF A PLASMA

A common gas such as air or hydrogen is made up of molecules which are electrically neutral. A molecule may consist of only one atom or it may be the union of two or more atoms. Ordinary hydrogen gas is diatomic (that is, its molecules consist of two atoms). Every hydrogen atom has a nucleus of one proton, normally accompanied by one electron, whose negative charge neutralizes the positive charge of the nucleus. A molecule of hydrogen, accordingly, has two atomic nuclei and two electrons.

The constituent molecules of a gas can be disrupted by subjecting them to ultraviolet light or X-rays, to an electric discharge, or to intense heat. What usually happens is that electrons are torn loose from molecules. The remnant of a disrupted molecule (an *ion*) fails by one electron to neutralize the positive charge of the nuclei and is therefore positively charged. We say the gas is *ionized*, which in our example means that some of the gas molecules have been split into positive ions and negative electrons (see Figure 6).

It is also possible to disrupt the molecules and free the atoms. If both the atoms are ionized, each molecule of hydrogen will be split into two free protons and two free electrons. Heavier

ATOMS

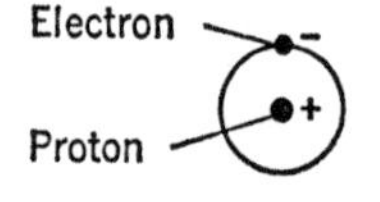

A *hydrogen atom* consists of a positive nucleus (proton) encircled by an electron. Total charge of atom = 0.

MOLECULES

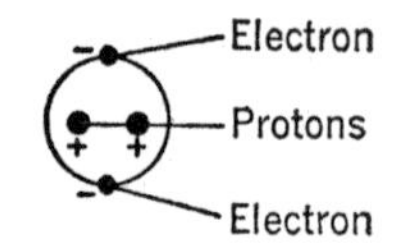

A *hydrogen molecule* is formed by two hydrogen atoms; consequently it consists of 2 protons encircled by 2 electrons.

IONIZATION

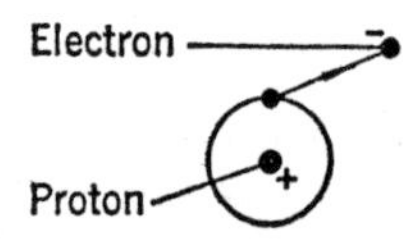

If the electron is torn away from the nucleus, the atom becomes *ionized*.

PLASMA

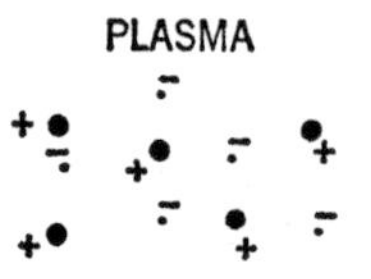

An ionized gas—for example, hydrogen—is called a *plasma*, and consists of free negative electrons and positive ions (= protons).

These diagrams show what happens in ordinary matter. In antimatter antiprotons take the place of protons, and positrons take the place of electrons.

AMBIPLASMA

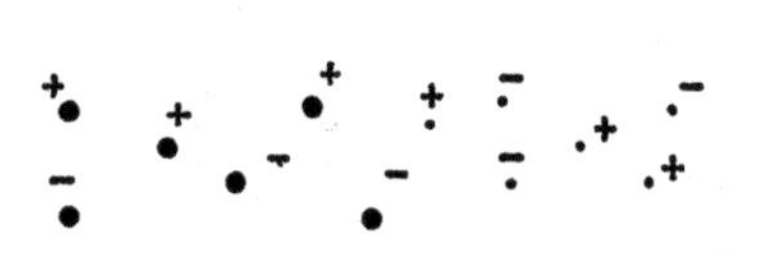

Ambiplasma is a mixture of protons and antiprotons, but it may also contain electrons and positrons. When protons and antiprotons collide, annihilation takes place. (Compare Figure 12.)

Figure 6. Atoms and molecules.

elements are ionized in a similar manner. A molecule of nitrogen, for instance, normally consisting of two nuclei (charge +7 and mass 14) surrounded by 14 electrons (2 × 7), can be split into two free electrons and two positive nitrogen ions. Each ion consists of an atomic nucleus surrounded by six electrons. Since these electrons only partially neutralize the seven positive charges of the nitrogen nucleus, the nitrogen has one net positive charge.

Ionization is counteracted by *recombination.* The positive ions attract electrons, which of course are negatively charged, and any electron that comes too close to an ion runs the risk of capture. The ion and electron are then recombined into a normal atom or molecule.

To employ the term that has gained increasing usage in the past twenty years or so, we call an ionized gas a *plasma.*

A plasma may be *completely ionized,* in which state all the molecules are divided into ions and electrons, or it may be *partially ionized,* with only a fraction of the molecules ionized and the remainder electrically neutral, normal molecules.

Generally speaking, a molecule becomes ionized if exposed to rough treatment, and it will recombine if given time to heal its wounds. Irradiation of a gas with ultraviolet light, X-rays, or gamma rays means that high-energy photons are shot through the gas. If they hit a molecule, ionization will often result. Alternatively, if a gas is heated, its molecules move with increasing speeds as the temperature rises; at a sufficiently high temperature the collisions become so violent as to disrupt the molecules—in other words, to ionize them. Under certain conditions this occurs at relatively "low" temperatures of between 5,000° and 10,000°C; under other conditions much higher temperatures are required. While ionization is in progress, ions and electrons are at the same time recombining. A state of equilibrium is achieved when the same number of molecules are recombining as are being ionized.

PLASMA IN COSMIC PHYSICS

In cosmic physics a dominating role is played by plasma. The stars consist of it entirely. Since the surface temperature of stars commonly ranges from 5,000° to 10,000°C, the plasma in the outer layers is often only partially ionized, but it is totally ionized in the hot interior.

Interstellar space, the space within the stars of a galaxy, is filled with a rarefied plasma. On the average there is only one atom in every cubic centimeter. Near a star, within a well-defined range designated by Bengt Strömgren as an HII region, the plasma is completely ionized by the star's high-energy radiation. But in the greater part of space, far away from any stars (HI regions), the ionization is only partial. The space between the galaxies (*intergalactic space*) is also filled with plasma, but at a much lower density. It is estimated that the density amounts to less than 10^{-6} atoms per cubic centimeter, or one atom per cubic meter.

If we consider that the stars are made up of plasma and that plasma occupies both interstellar and intergalactic space, it becomes evident that most of the universe consists of plasma. But interstellar and intergalactic plasma, with its very low density, differs appreciably from the dense plasma of the stars. Since plasma in space is of chief concern to us here, we shall focus our discussion upon its properties.

MAGNETIZED PLASMA

A plasma is influenced by magnetic fields. Although such fields in space are admittedly feeble, they are strong enough, given the low density of the plasma, to exert decisive effect on its properties. The usual intensity of a magnetic field in interstellar space is believed to be 10^{-5} or 10^{-6} gauss, or approximately one hundred thousandth that of the earth's magnetic field (which is 0.3–0.6

gauss). The intensity is presumably much less in intergalactic space.

A magnetic field affects the motions of electrically charged particles (see Figure 7). In the plasma of space the motions of both electrons and ions are essentially determined by the magnetic field. But the neutral (normal) atoms and molecules are not affected. They generally move in a straight line when remote from the stars; in the vicinity of a star they yield to gravitation and describe the orbit of a falling body. In the course of their motions they collide with other atoms or with ions or electrons, causing them to change speed and direction. The average distance they traverse between collisions is called their "mean free path" and depends on the plasma's density. If the density is one particle per cubic centimeter—a characteristic value for interstellar space—the mean free path comes to about 10^{15} cm or one thousandth of a light-year, which approximates the diameter of

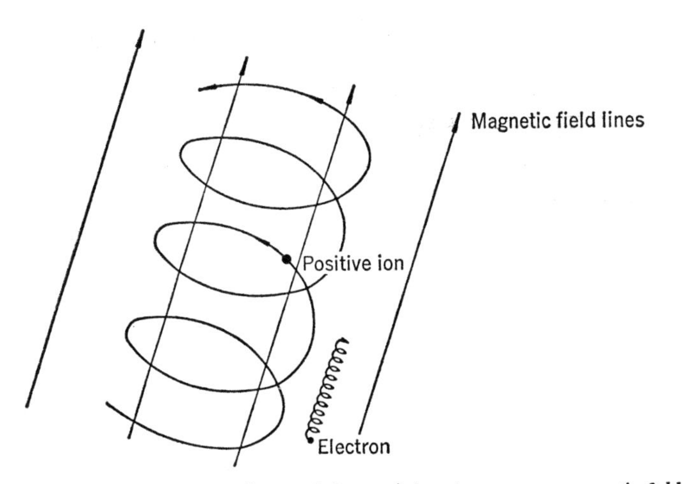

Figure 7. Charged particles (ions and electrons) in a homogeneous magnetic field. The particles move in spirals, those of the electrons and those of the positive ions being in opposite directions. The spirals of the electrons are much thinner than those of the ions (if both kinds of particles have the same energy). The more energetic a particle, the thicker the spiral.

Pluto's orbit (and hence the dimensions of our planetary system).

In the absence of magnetic fields the ions and electrons would generally resemble the atoms in their motions. But a magnetic field acts like a rudder in controlling the motion of a charged particle (with the axis of the tiller turned in the direction of the magnetic field); depending on the charge, the rudder will be locked to port or to starboard. If the particle moves at right angles to the magnetic field, it will describe a circle. If at the same time it also moves parallel to the field, this motion will not be affected by the field. The result will then be for the particle to move in a spiral of the same type as that of an airplane which, say, is bearing continually toward the left, but at the same time is gaining or losing altitude. The thickness (or diameter) of the spiral derives from the particle's energy and the strength of the magnetic field.

OUTER SPACE IS BOTH COLD AND HOT

An artificial satellite traveling close to earth is heated by the sun's rays, yet at the same time it is being cooled by the process of radiating away its own heat. The balance between received and emitted heat results in a temperature similar to the earth's mean surface temperature (which, after all, is determined in the same manner). This temperature is called the radiation temperature (though to hold strictly true in our example, the satellite must be what a physicist would call a "black body"). A different situation emerges if we visualize a body traveling in the vast darkness beyond our planetary system: it will receive very little heat from the stars, but it will still be emitting its own heat. Gradually, its temperature will come fairly close to absolute zero, −273°C. The radiation temperature of outer space is very low, so that any solid body there will assume a temperature close to −273°C (see Figure 8).

Outer space is hot, however, in the sense that the particles

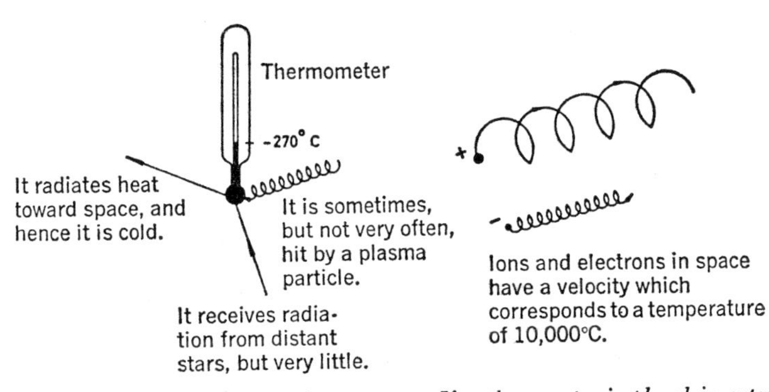

Figure 8. The temperature of outer space. If a thermometer is placed in outer space, it shows a very low temperature. But the very thin gas in outer space (consisting of ions and electrons) is very hot.

constituting the plasma are in a state of agitation that corresponds to a high temperature. This is because the particles in thinned-out plasma gain energy by absorbing the radiation of the sun, but lack the ability of a solid body to radiate energy. Thus in an HII region, close to a star, the plasma has about the same temperature as the star (perhaps from 5,000° to 10,000°C). In the HI regions, far from the stars, the plasma is cooler (perhaps only a few hundred degrees above absolute zero), though still much warmer than the temperature of a solid body (the radiation temperature). It might be thought strange for a solid body to be cool in the middle of hot space. An appropriate analogy here is an ice-cream cone, which takes a long time to melt even on a hot summer day. We must remember that the plasma of outer space has a density only 10^{-19} that of air, better than any vacuum that can be produced on earth; hence the solid body heats up only negligibly even if the plasma is hot.

THE MOTIONS OF PLASMA PARTICLES

In a plasma, as in a solid body, the particles (atoms, electrons, ions, and so on) that it comprises are always in motion. As the temperature rises, the *thermal motion* increases. In a magnetized plasma of the type that fills space, the temperature determines the thickness of the spiral which a particle follows in its motion. The greater the temperature, the more extended the spiral. As the strength of a magnetic field increases, however, the spiral contracts.

If a part of space contains a magnetic field of 10^{-5} gauss (a reasonable value according to the above) and if the particles there have a temperature of 10,000°C, the electrons move in spirals having a thickness of about 6×10^5 cm, or only 6 km, which by cosmic standards is very small. The spirals of the protons move in the opposite direction and their thickness is 2.5×10^7 cm, or 250 km—still insignificant by the cosmic yardstick. (If the temperature is only 100°C, these figures are reduced to one-tenth.) But if we compare these distances with that traversed by neutral atoms between collisions (10^{15} cm), we find that the magnetic field, though only as feeble as 10^{-5} gauss, has reduced the freedom of motion of the charged particles by a factor on the order of millions or billions of times.

The foregoing conclusion, it should be noted, applies only to the motion of particles at right angles to the magnetic field. In a direction parallel to the field the motion may continue uninhibited at a uniform velocity, provided the field intensity and direction remain unchanged. In that case a particle might move, say, 10^{15} cm in the field's direction before its motion is changed by collision with another particle. It is highly improbable, however, that magnetic fields in space behave with anything like the regularity assumed here. As a particle spirals its way along a magnetic line of force it will encounter fields which become stronger or weaker. The spiral's course will consequently change. If a particle enters a region with a sufficiently strong magnetic field, the spiral may reverse, propelling the particle in an opposite direction. A magnetic field of varying intensity and direction therefore reflects many of the spiraling particles.

Assuming that a plasma is in thermal equilibrium, with 100 of its particles spiraling toward a stronger magnetic field, only about five of them can reach a point where the field is 10 times stronger. The other 95 will be reflected before then. It is likely that the general magnetic fields in space are "mirrors"—that is, fields, which in this way reflect the particles. As a result, the particles will often oscillate back and forth between two turning points in their spiral orbits (see Figure 9). Thus their freedom of motion

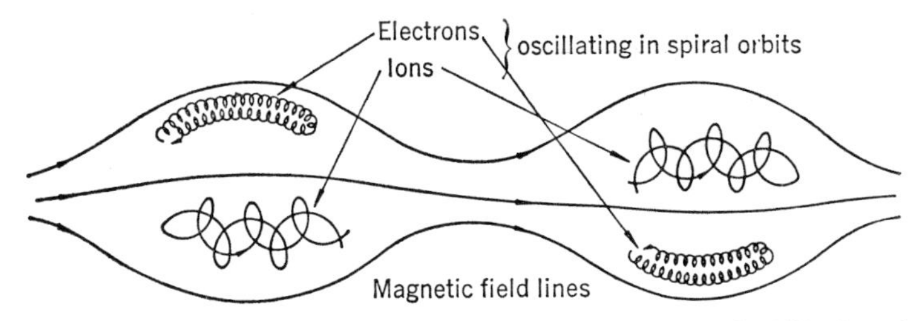

Figure 9. Charged particles in space. In a homogeneous magnetic field, charged particles move in spirals with a constant speed (see Figure 7). As the strength of the magnetic field varies in space, the particles will usually oscillate to and fro.

will be greatly reduced even in the direction of the magnetic field. Just how much it is reduced will depend on the field's irregularity, about which we know little.

Summing up the result of our analysis, a magnetic field of the intensity we can expect in interstellar space greatly reduces the motions of charged particles. By cosmic standards, taking one light-year as our unit of measure (10^{18} cm), their mobility is very limited indeed. They move in tightly-wound spirals (on the order of less than one-billionth of a light-year) and with probably a short length as well (perhaps one-millionth of a light-year).

As for intergalactic space, our knowledge is considerably less, but the same general conclusions ought to hold. A suitable galactic yardstick would be 10^{22} cm or 10,000 light-years, in relation to which the mobility of charged particles fades to insignificance.

HIGH-ENERGY PARTICLES. COSMIC RADIATION

So far we have been considering the motion of electrons and ions at thermal speeds or the speeds induced in these particles by heat. We have seen that such particles cannot easily travel great distances in space. But what about the so-called high-energy particles?

The energy of a particle (electron or ion) is often measured in electron volts. This is the amount of energy gained by a particle (with unit charge) when accelerated by a difference in electric potential of one volt. If we have a high-tension plant of 2 million volts, we can use it to accelerate electrons to an energy of 2 million electron volts (2×10^6 or 2 MeV). The modern giant accelerators in Geneva, Berkeley, Brookhaven, and Dubna are capable of generating 30 billion electron volts (3×10^{10} or 30 BeV). Particles with energies as enormous as 10^{19} or 10^{20} electron volts have been detected in cosmic rays, but these are very rare. The cosmic radiation we are most familiar with contains particle energies of between 10^9 and 10^{11} electron volts, and the majority of particles happen to fall within this range.

If we were to shoot out charged, high-energy particles from the sun or some other part of the solar system, they too would move in spirals; the higher the energy of any one particle, the more extended would be its spiral. The pathway that the ejected particles followed would resemble a slender hose stretched out from the solar system along the magnetic lines of force. The particles could hit a certain star only if the star was in, or at least near their pathway, but their chances of doing so are very small so long as the path is narrow. It is only when the path has a width of, say, one light-year that the particles stand a reasonable chance of hitting the star. If we figure out the particle energy this would require, we come to a figure of 10^{14} electron volts. This is 3,000 times more than has been attained with the biggest accelerators. Unless we can provide enormous energies of this magnitude, our chances of aiming electrons, protons, or ions at the vicinity of a star are rather poor.

The magnetized plasma of interstellar space thus acts as an effective barrier to the passage of elementary and atomic particles from one star to another. It is too thin, however, to impede the motion of larger bodies. The course of a spaceship or even of a small grain of sand would not be noticeably affected by interstellar plasma. Plasma and magnetic fields don't begin to make themselves felt until a traveling object is less than 1/100 mm in size.

OUTER SPACE IS BOTH EMPTY AND VISCOUS

By now it should be apparent that space between the stars is not the empty, desolate place that has been commonly imagined. The commander of a spaceship might safely regard interstellar space as empty in the sense that its matter is so thin and its magnetic fields so weak as not to affect his navigation noticeably. Nor would there be any effect on smaller bodies—even on grains of sand so tiny that the naked eye can barely see them.

But electrons, protons, and ions (as well as microscopic grains of dust) are strongly influenced by plasma and magnetic fields. As far as they are concerned, space is a viscous medium, almost like syrup and just as sticky. Once caught in it, they normally have the flimsiest of chances for getting out, much less for moving from one star to another.

In all probability it is normal for a star to lie embedded in a thick layer of plasma, which often may be one light-year across. Part of the plasma may have been ejected from the star itself. Or the star may also be regarded as a part of the plasma, which at one time condensed under the influence of gravitation. In either case there must surely be the closest association between a star and the immediately surrounding interstellar plasma.

A similar kind of interaction may be assumed between a galaxy and the plasma surrounding it in intergalactic space.

V

Antimatter in the Cosmos

KOINOPLASMA AND ANTIPLASMA IN SPACE

After the introduction to plasma physics in Chapter IV, we are better prepared to reconsider the question: Does the universe contain antistars, stars made up of antimatter?

Because of the strong interaction between a star and the plasma immediately around it, it follows that the two should contain the same kind of matter. We say "should" because we cannot rule out the possibility that a star of koinomatter has chanced to enter a space containing antiplasma, and vice versa. But the more normal pattern is likely to be that a star has plasma of the same kind in its surroundings.

It then becomes important to find out what happens if two adjacent stars consist of different matter. Suppose that a koino-star and an antistar are separated by, say, four light-years, and each is surrounded by plasma of the same kind that it has. At some point in the intervening space there must be contact between an extended koinoplasma and an extended antiplasma. What phenomena arise when they intermix? The obvious answer is an annihilation, which generates such intense energy as to heat up the plasma in the border region. To understand what then happens, we shall first study a very simple and well-known phenomenon.

THE LEIDENFROST PHENOMENON

The following demonstration can be performed in an ordinary kitchen. Heat a metal plate and put a drop of water on it. An electric hot plate will do nicely, especially if it is of the type with a hollow in the center. If it has, put the drop in the hollow.

At a temperature slightly above 100°C, the drop of water will evaporate almost immediately, making a loud hissing noise as it does so. When the temperature is raised a bit more, the drop vanishes in explosive fashion in the fraction of a second. But a temperature of several hundred degrees, enough to make the plate glow red, causes another phenomenon. The drop does not boil off immediately. It can be made to stay quietly in place for more than five minutes, though it will oscillate to and fro. Then it gradually gets smaller until it disappears entirely. But if the plate's temperature is suddenly lowered while the drop is still there, an explosion occurs and the drop vanishes.

The phenomenon is named after a German physician, Dr. Leidenfrost, who studied it in the 19th century. What happens is that, as the drop evaporates, a layer of steam is formed between the drop and the hot plate (see Figure 10). This layer tends to *insulate* the drop from the plate, so that the latter's heat is conveyed to the drop more slowly. If one wishes the drop to evaporate slowly, then a sufficiently thick insulating layer of steam requires exceedingly high plate temperature. At a plate temperature only slightly above 100°C, the steam layer is too thin to insulate the drop, and the drop goes off in a puff of vapor.

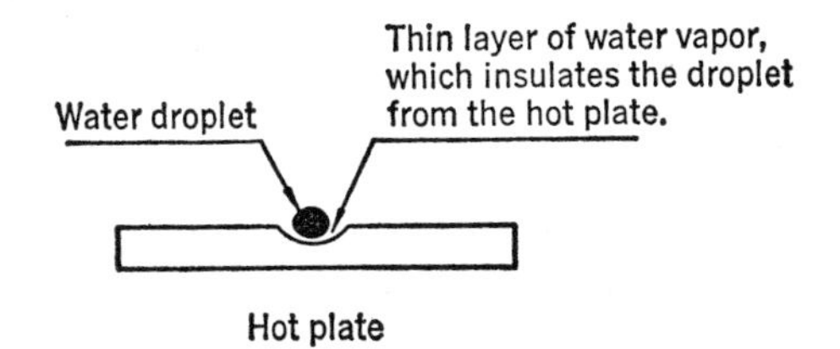

Figure 10. The Leidenfrost phenomenon.

Let us now repeat the experiment, this time replacing the drop with a bit of antimatter. Even if we happen to have the bit at home, the kitchen will not do, which is unfortunate (or should we say fortunate?). But if in our hypothetical experiment we put a bit of antimatter on a plate of koinomatter (in a koino-kitchen!), a similar phenomenon might occur. The first contact between koinomatter and antimatter will result in annihilation, but the attendant generation of energy will create a force that separates the two. In the Leidenfrost phenomenon, as we noted, a layer of steam is formed between the drop and the plate. In our more ambitious experiment, a layer is similarly formed to separate the koinomatter and antimatter. Reasoning by analogy, we may expect this to bring about an annihilation that is relatively slow and small in scale. It need not be more intensive than to ensure the formation and maintenance of an adequate insulating layer.

Let us try to apply these lines of thought to our cosmic problem. Suppose that an ordinary koinostar and an antistar are adjacent and that each is surrounded by interstellar plasma of its own kind. We must then look for a surface of contact between the koinoplasma and antiplasma somewhere in the intervening space, since it is there that we get our analogy to the Leidenfrost phenomenon. When the koinoplasma and antiplasma intermix, annihilation ensues, accompanied by generation of enormous energy. As the border layer heats to a very high temperature, it becomes highly attenuated. The annihilation in this layer may then proceed very slowly, because the rate of annihilation need not be more than necessary to maintain the layer. In due course a "Leidenfrost layer" may form to insulate the koinoplasma from the antiplasma (see Figure 11).

Calculations show that a Leidenfrost layer in space can be very thin by cosmic standards. Under certain conditions it may not be more than 1/1000 of a light-year across, and may possibly be less. That would seem to be enough to keep the koinomatter and antimatter effectively separated from one another.

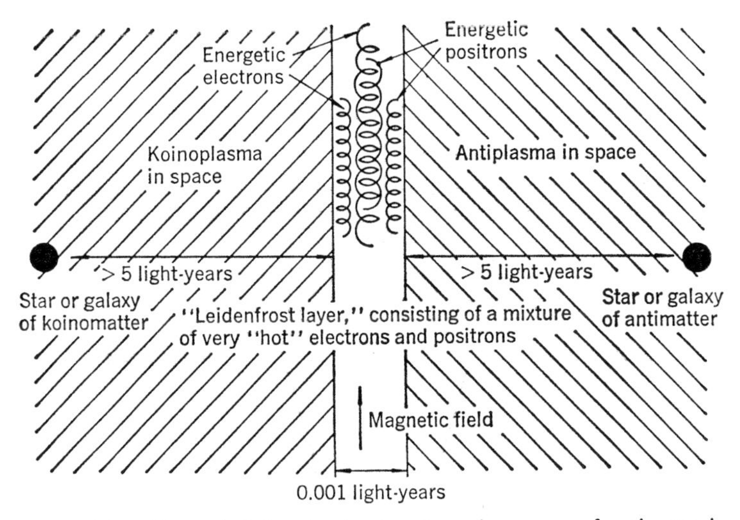

Figure 11. The annihilation sheath separating koinomatter and antimatter in space.

This suggests our next question: Assuming the existence of two neighboring regions of antimatter and koinomatter in space, how do we prove the existence of a separating Leidenfrost layer? To put the problem in these terms is to imply a more encompassing problem: to derive the general properties of a plasma consisting of intermixed koinomatter and antimatter.

AMBIPLASMA

Here we shall use the term *ambiplasma* to describe a mixture of koinomatter and antimatter. The word is derived from the Latin "ambi," meaning *both* (as in ambivalent, ambidextrous, and so on).

Let us assume that one region of space contains a plasma of mixed protons and antiprotons. We assume that this plasma, like all other plasmas in the cosmos, is magnetized. Our concern here is to study how such an ambiplasma behaves.

An antiproton will occasionally collide with a proton, resulting in annihilation (see Figure 12). Upon collision, the two particles

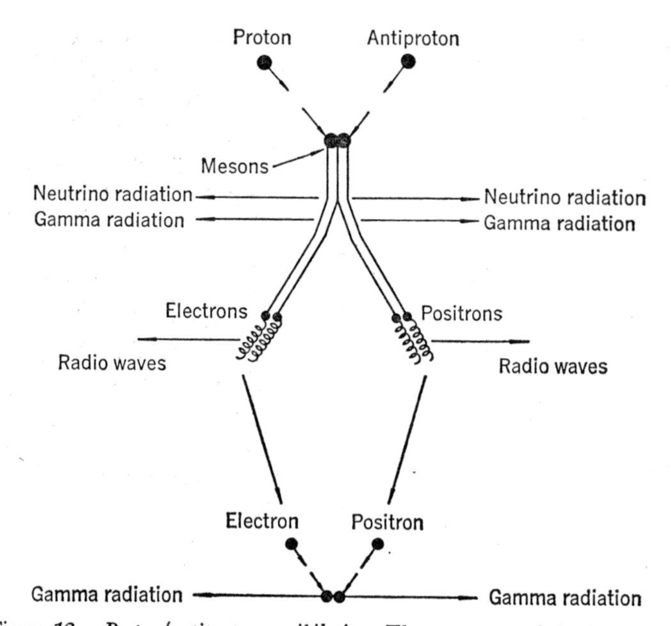

Figure 12. Proton/antiproton annihilation. The upper part of the figure shows in a simplified way how a collision of a proton and an antiproton produces mesons, which rapidly decay and emit neutrinos and gamma radiation. After a few microseconds there remain fast electrons and positrons, spiraling in the magnetic field and emitting radio waves. The lower part of the figure shows that if one of these electrons collides with a positron, an annihilation takes place, resulting in the emission of gamma rays.

will first transform into mesons, which are hurled outward. They disintegrate spontaneously into other mesons and finally into electrons and positrons. Only a few microseconds after the collision we have obtained one or two electrons and one or two positrons, with energies of about 100 MeV (10^8 electron volts) each. These particles stay close to the collision site, since the magnetic field prevents them from going astray. There has also been production of gamma rays and neutrino rays, which leave the collision site with the speed of light—they are not captured by the magnetic field.

If we compose a balance sheet of energy for the collision

account, the credit sidc shows the energy equivalent to the masses of the proton and antiproton. On an energy scale a proton mass reads at 900 MeV and an antiproton mass at the same figure. If we assume negligible motion by the particles prior to collision, we have a surplus of 1,800 MeV. Of this quantity, about 900 MeV is generated as neutrino radiation, which is extremely difficult to detect, since neutrinos interact very little with other particles. Our accounting should therefore attribute the entire neutrino energy generation as a loss. We deduct another 600 MeV to account for the gamma rays, which may be measured by scintillators and other detectors common in nuclear physics. Gamma radiation may be regarded as a sure indicator to demonstrate the occurrence of annihilation.

Our last entry is for the electrons and the positrons. The processes of annihilation may take different courses, but the average end result is 1.5 electrons and 1.5 positrons for each proton-antiproton annihilation. (Of course we really mean to say that, for every *two* proton-antiproton annihilations there are, on the average, *three* electron-positron pairs formed.) Since the electron's mass comes to 0.5 MeV in terms of energy, the trio's mass energy totals 1.5 MeV. Their combined kinetic energy, however, is close to 300 MeV. The account now reads:

Table 3. Balance Sheet of Energy

Initial Energy	
Mass of proton	900
Mass of antiproton	900
Total initial energy	1,800
Resulting Energy	
Neutrino rays	900
Gamma rays	600
Electron and positron energy (the mass energy is only 1.5 MeV; the rest is kinetic energy)	300
Total resulting energy	1,800

With the neutrino and gamma rays out of the way, the high-speed electrons and positrons are the only actors left on the stage. They move in spirals of the kind already described, but since their energy is much higher than the relatively low-energy particles we talked about earlier, their spirals will be much more extended. In a field of 10^{-5} gauss, their diameters will approximate 10^{13} cm, somewhat less than the distance between sun and earth. However, this is only 10^{-5} (1/100,000) of a light-year, which is very small by the cosmic yardstick.

As more and more protons and antiprotons are annihilated, a plasma is produced consisting of extremely high-energy electrons and positrons which spiral in the cosmic magnetic field. So high is their energy that it is equivalent to the fantastic temperature of 10^{12} (one trillion) degrees. Thus if an ambiplasma originally consists of only protons and antiprotons, an extremely hot electron-positron gas is automatically generated. The high temperature gives a fair idea of the enormous energy released.

If we heat a gas or a plasma, its pressure increases and it tries to expand. If a koinoplasma and an antiplasma adjoin in some part of space, they will intermix in the border layer to form an ambiplasma. Annihilation in the border layer gives rise to a tremendous increase in temperature, causing the layer to expand. As a result, the antiplasma and koinoplasma repel one another, making further contact between them difficult or impossible. We have here our analogy to the drop of water on the hot plate, where a layer of steam intervenes between the two.

When the annihilation of protons and antiprotons in an ambiplasma has proceeded far enough to give rise to a large number of electrons and positrons, it becomes increasingly probable that these will collide. When that stage is reached, the annihilation of electrons and positrons also takes on importance. What happens is that all their energy, both of mass and motion, is emitted as gamma rays. When all the electron-positron pairs created by the proton-antiproton annihilation have thus snuffed out one another, we reach the end of the annihilation process. All the mass energy of the original protons and antiprotons has now been transformed into radiation.

EMISSION OF RADIO WAVES FROM AN AMBIPLASMA

If we wish to send out radio waves from a radio or TV station, we cause an electric charge to oscillate in the station's antenna. Indeed, any oscillation of an electric charge will cause electromagnetic radiation, of which radio waves are only one form. Electrons and positrons spiraling in their orbits within a magnetic field obviously represent oscillating electric charges. They must therefore emit radio waves. This method of producing radio-wave emission is often called synchrotron radiation, because it first demonstrated its importance in the type of particle accelerators known as synchrotrons. The energy emitted is taken from the kinetic energies of the electrons and positrons.

The electrons and positrons produced in a magnetized ambiplasma thus emit radio waves, whose wavelength depends on the strength of the magnetic field and the energy of the particles. Emissions from a cosmic ambiplasma will lie in what radio engineers call the shortwave band.

Under certain conditions electrons and positrons may lose much of their kinetic energy through emission of radio waves. This, it should be noted, is in addition to the neutrino rays and gamma rays already lost during proton-antiproton annihilation. When the conditions are favorable, almost all the kinetic energy (totaling 300 MeV according to Table 3) is emitted as radio waves.

The exact apportionment between gamma rays and radio waves will depend on how fast the different processes occur, which in turn depends on the density of the ambiplasma and the strength of the magnetic field. The greater the plasma density, the faster the annihilation; the stronger the magnetic field, the faster the radio radiation. If plasma density is very low and the magnetic field is very strong, the electrons and positrons will emit almost their entire energy as radio waves. Under opposite conditions (high density, weak magnetic field) virtually all the energy is emitted as gamma rays and neutrino rays.

EVIDENCE OF COSMIC AMBIPLASMAS

Given the conditions assumed in space, about half the energy released in the annihilation may be expected to go out as neutrino rays, one-third as gamma rays, and one-sixth as radio waves.

Neutrino rays are very difficult to detect. In any case, the measuring instruments we have today certainly do not react to neutrinos from faraway cosmic sources.

To detect and measure gamma rays we rely on scintillation crystals. The gamma rays are not measured directly: instead, they must first dislodge electrons ("secondary electrons") from the atoms of the matter through which they pass. When these electrons pass through the crystal, they produce photons which are collected by a photomultiplier, and here they dislodge new electrons. It is these new electrons that are led to the electrical measuring apparatus. For the quantitative determination of energy this complex process is very inefficient, since no more than a small fraction of gamma-ray energy is detected by the apparatus. The apparatus, moreover, is usually rather small: the cross section which captures the gamma radiation is often of the order of one square inch.

Radio waves are much easier to detect. The radio telescope, which measures such waves coming from space, focuses them with enormous reflectors or with aerial systems that may cover several square miles. Thanks to an exceptionally refined detection apparatus, a large part of the collected energy can be measured.

For all these reasons, cosmic radio waves are much easier to detect than cosmic gamma rays. The faintest detectable emission of gamma rays must still transfer 10^8 (100 million) times more energy than the faintest detectable emission of radio waves. In other words, radio engineers are 100 million times more proficient than nuclear physicists when it comes to measuring radiated energy.

We may conclude that, even if an ambiplasma in space emits twice as much energy in the form of gamma rays as in the form of radio waves, the latter are by far the easier to detect. Our nuclear instruments don't give us any chance to detect gamma radiation until the radio telescopes register 10^8 times the threshold level for radio waves. (And the signals would of course have to be many, many times stronger to permit the detection of neutrino rays.) Even on the extreme assumption that only 10^{-6} (one part in a million) of the energy from an ambiplasma is emitted as radio waves, the existence of such waves would still be 100 times easier to demonstrate, compared with gamma rays.

A cosmic ambiplasma must therefore primarily reveal itself to us through the emission of radio waves. Taking the argument a step further, it is to the *radio stars* that we must first look when searching for an ambiplasma in cosmos. Such stars might consist of ambiplasma, which emits radio waves. But it must be remembered that an emission of radio waves may take place in many ways, so that ambiplasma is not necessarily the means of emission from radio stars. Or, again, there may well be ambiplasma in space with radio emission so small as to escape detection.

Earlier we discussed the properties of a Leidenfrost layer in space, capable of separating an area of koinomatter from an area of antimatter. Since by definition the border layer contains ambiplasma, it must emit electromagnetic radiation, including radio waves. It is very interesting to venture an estimate of the radiation strength, because that could help us assess our chances of detecting such a layer, assuming it exists in the first place. A calculation shows that a stable Leidenfrost layer will not emit so much energy that we can expect to detect it by our present instruments. This is the same as saying that space may very well contain layers that separate koinomatter from antimatter, yet we have no chance to detect them by measuring their radiation. The statement assumes, however, that such layers would be fairly thin and relatively *undisturbed.*

But there may be areas in space where koinomatter and antimatter are *intermixed* to the point of generating intensive annihilation and powerful emission of radio waves. Thus a border zone between koinomatter and antimatter need not give rise to a radio star, but a radio star might very well arise from intermixture in the zone.

CONCLUSIONS ON THE EXISTENCE OF ANTIMATTER

In a former day, when the electron had been discovered but not yet the positron, scientific theory—with not a little support from esthetic and philosophical circles—insisted on symmetry between positive and negative charges. This demand required the existence of a positive electron, but no such particle could be proved by experiments. There were only two ways to solve this conflict of principle: either the theoreticians give up symmetry as a lost cause or else the experimenters discover a positive electron. As things developed, the demand for symmetry proved sturdier than the then known body of experimental facts. The discovery of the positron in 1932 was a triumph for the esthetic-philosophical aspect of physics.

A similar term of reference was applied in the ensuing quarter-century, but now the focus of attention was on the heavy particle, the proton. When the antiproton was finally discovered, the criterion of symmetry had shown its strength for the second time.

We confront a like situation for the third time, but now attention has been shifted from the physics of elementary particles to cosmology. There is no definite evidence for the existence of antimatter in the cosmos, but we are pretty well familiar with the properties of matter. The physics of elementary particles tells us that positrons and antiprotons "exist" in the sense that they can be produced in accelerators. Their properties, we know, are completely symmetrical with those of the electron and proton. With improved experimental prowess, we know how we could go about producing antimatter possessing all the complex properties of koinomatter; we could, in short, produce a complete "mirror world" or "antiworld" of antimatter. Assuming that antimatter exists in some part of the universe, we feel convinced that it would contain a world similar to our own. However, we have no proof that antimatter exists in reality.

If we hold that the laws of physics apply to the structure of the universe generally, then the requirement of cosmic symmetry between koinomatter and antimatter is bound to assert itself very strongly. After all, symmetry's twofold triumph in discovering the positron and antiproton is highly suggestive of its validity. We must not forget, however, that there is a vast difference between particle physics and cosmology.

Indeed, we accord so much weight to the symmetry criterion that we think it worth the trouble to review the arguments for and against the existence of antimatter. To begin with, we have found no *crucial* argument for its existence. A mixture of koinomatter and antimatter should manifest itself by emission of radio waves, and so the radio stars may indicate the presence of antimatter. (It is also possible that the radiation of radio stars is amenable to other explanations.) But neither have we found any argument *against* the existence of antimatter. However, we can satisfy the demand for symmetry in the universe by following various paths of supposition.

1. We may suppose that antimatter is present in some remote corner of space unknown to us and that the whole of our metagalaxy consists of koinomatter.

Alternative: The metagalaxy is symmetrical, with half of it consisting of antimatter and the other half of koinomatter. However, this condition can be satisfied in different ways.

2. We may suppose that every second galaxy is solely made up of antimatter and every second galaxy of koinomatter.

Alternative: The symmetry criterion could be fulfilled within each galaxy. Our own galaxy would then be equally divided between koinomatter and antimatter. This division can happen in any of various ways.

3. We may suppose that the remotest half of our galaxy consists of antimatter and the nearest half is of koinomatter.

4. We may suppose that every second star in our vicinity consists of antimatter. This is carrying the symmetry requirement to its extreme.

Perhaps the most shocking result of our analysis is that we cannot with certainty even rule out Supposition 4 at the present time. If someone were to claim that Sirius, the brightest fixed star in our firmament, consists of antimatter and not koinomatter, we would not have any tenable argument to demolish his claim. Were Sirius to consist of antimatter, it would have exactly the

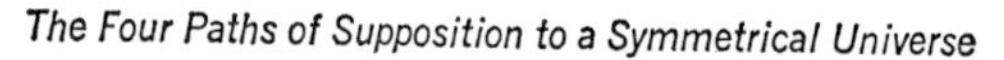
The Four Paths of Supposition to a Symmetrical Universe

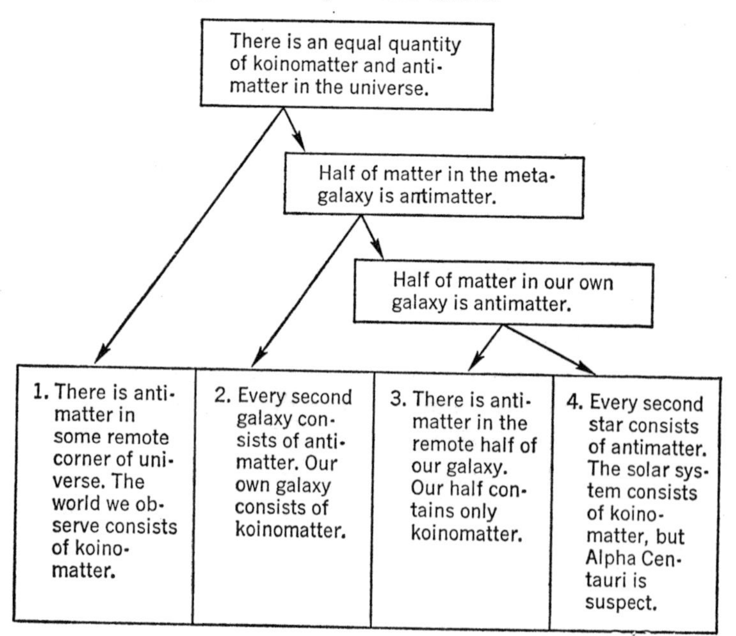

same appearance and emit the very same spectrum as if it consisted of koinomatter. The space around Sirius must then contain antimatter, but this may be separated from the koinomatter in the space around us by a thin Leidenfrost layer, which we are poorly equipped to detect.

We must therefore leave unanswered the question of the existence of antimatter. But in view of the compelling arguments for symmetry and the lack of evidence against it, we are motivated to pursue our investigations of antimatter in the cosmos. In the next chapter, therefore, we return to the cosmological speculations with which we began this book.

VI

Development of the Metagalaxy

NATURAL LAWS AND COSMOLOGY

The cosmological problems were first discussed in Chapters I and II, and it was pointed out that they can be approached in several ways: the mythological and the purely scientific, with any number of hybrids in between. Which approach one prefers is largely a matter of taste, conditioned by one's general attitude to philosophical and religious questions. So far as a sound natural science is concerned, however, a distinct separation from mythology is a matter of life and death.

It is one of the essential tasks of the scientist to explain how far the borders of our knowledge extend. One advance toward drawing such a boundary line is to contemplate all observed phenomena in terms of the natural laws discovered in the laboratory. In the field of astrophysics, we can phrase our search in these terms: Have we observed any phenomena that require the introduction of new natural laws?

This raises the question as to whether natural laws or physical constants are dependent upon time or space (already discussed in Chapter II). Can we assume that a measurement or physical experiment made in some remote part of the universe at a remote point of time will yield another result than one made here and now? (We assume, of course, that our measurement has been corrected for the state of motion of the measuring system and the gravitational field in which the measurement is carried out.) We have no reason to believe that this is the case. So far we have found no astrophysical phenomenon that compels us to formulate new natural laws. Nor is there any reason to believe that natural laws were different at an earlier time, or that other laws apply elsewhere in the universe.

Obviously, we cannot rule out the possible existence of natural laws unknown to us. But the task that has meaning for us is to explain as much as we can in terms of the laws we already know. In other words, we seek to bring the maximum number of astrophysical phenomena into harmony with our everyday existence—more specifically, the everyday existence of laboratory physics. This of course includes quantum theory and relativity theory.

Since the discovery of the positron and antiproton, the "everyday world of physics" also embraces antimatter. By postulating with Professor Klein the symmetry of matter-antimatter throughout the world, we are not sneaking in an ad hoc assumption. We are merely applying the everyday world of elementary-particle physics to astrophysical phenomena, as has already been done for mechanics, electrodynamics, and nuclear physics.

In doing so we question the long-standing assumption that all celestial bodies consist of koinomatter. It was a natural assumption to make before the discovery of the antiparticles, but it is old hat today.

We shall therefore base the following analysis on two principles.

1. There are no new natural laws.
2. There is symmetry between koinomatter and antimatter.

Strictly speaking, the analysis contained in Chapter II, before we began to discuss antimatter, does not conflict with these principles. That is because we can get around the symmetry requirement by assuming that the whole world we have observed up to now consists of koinomatter, and that antimatter exists in another part of the world which we haven't yet discovered or which is theoretically unobservable (our first "path of supposition," page 62). But this is a rather artificial assumption. It immediately begs the question: How did the koinomatter and antimatter become separated? The only plausible answer is to assume a primordial state in which there was separation from the beginning.

A more attractive assumption, at least from certain points of view, is that koinomatter and antimatter were homogeneously blended in the primordial state. To answer the nagging question, "How did that happen?" we can invoke the radiation materialized in conformity with processes familiar to us (Chapter II). We know that sufficient high-energy radiation can produce protons and antiprotons, in equal amount per unit of volume. Electrons and positrons are likewise produced in pairs. We have actually made no assumption other than the existence of a universe containing energy in a form that can be materialized in accordance with known natural laws. Our assumption of a primordial state is the next simplest one we can make. The simplest is an empty space without energy. But this is uninteresting. *Ex nihilo nihil.*

IN THE BEGINNING, AMBIPLASMA

We find ourselves at the limits of our knowledge and do not intend to penetrate the philosophical arguments to greater depth. Instead, we start our discussion by assuming a primordial state with a homogeneous mixture of koinomatter and antimatter, giving us what we call an ambiplasma. Protons-antiprotons and electrons-positrons may be assumed to have been present in the primordial state. To make the assumption more manageable, let us say that at first only protons and antiprotons exist. From these, electrons and positrons are automatically formed by annihilation. Whichever assumption we use, we eventually get both heavy and light particles.

Now let us see how the ambiplasma develops under the influence of ordinary natural laws. The assumption we have made is difficult to reconcile with the big-bang theory, which postulates extremely *high* density as its starting point. But if we have an ambiplasma of high density we get a very rapid annihilation. An ylem consisting of ambiplasma is annihilated and instantaneously converted into neutrino rays and gamma rays. But it would appear to be difficult, if not impossible, to arrive at our present world in this fashion. The big bang would be too big a bang!

We are therefore forced to start with an ambiplasma having very *low* density. This ambiplasma is premised to fill all of the volume we can perceive. (For the moment we leave indefinite what lies beyond, but we shall deal with this question later.)

For purposes of theoretical analysis it is important to consider that the ambiplasma is magnetized. This is a reasonable assumption, since so far as we know all cosmic plasma is magnetized. We shall disregard the question of whether magnetization must be introduced as an extra assumption in regard to the primordial state or whether processes exist which automatically induce magnetization.

Theoretically the primordial ambiplasma could have been distributed uniformly throughout its volume, or else the ambiplasma could have been more dense in some regions than in others. If the first was the true, original situation, then inevitable statistical fluctuations would have ensued which led to the second situation. Thereafter, the regions of greater density gave rise to gravitational forces that tended to make the density variations even more pronounced.

Under the very general conditions we have assumed, it becomes difficult to study what is happening. A clearer picture will be obtained if we backtrack and analyze a highly simplified *model*. The assumptions we now introduce are not of fundamental importance; their only purpose is to simplify our treatment and make it easier to comprehend.

A METAGALACTIC MODEL

The starting point in our model is an ambiplasma which fills a huge sphere (see Figure 13a). Its density is uniform throughout. Assume the sphere has a radius of one trillion (10^{12}) light-years and the density is so low that, on an average, there is only one proton or antiproton in every cube with sides of a hundred meters. The probability that a proton will collide with an antiproton in such a void is next to zero, which means that annihilation is negligible. The only force of importance acting on the ambiplasma is gravitation. With every part of the sphere attracting all the other parts, the sphere begins to contract. Over trillions of years the radius of the sphere will decrease and its density increase.

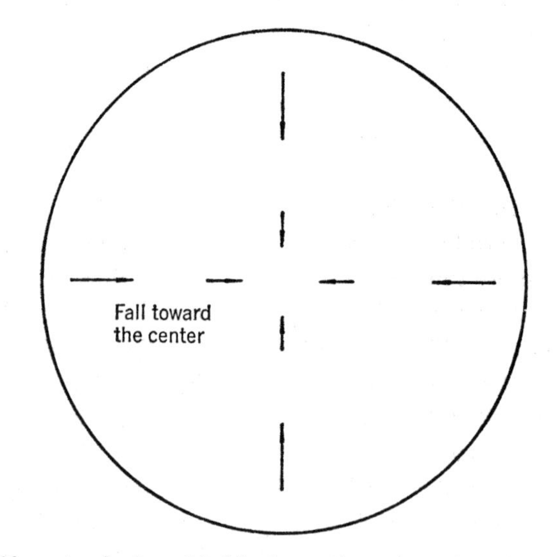

Figure 13a. A galactic model. (1) A very big sphere of ambiplasma contracts under the action of gravitation.

According to the law of gravity, two bodies attract one another with a force proportional to the product of their masses and inversely proportional to the square of the distance between them. That part of the mass in the center of our huge sphere will be attracted by all the surrounding parts. But since these forces are symmetrically aimed in all directions, they cancel out one another. The resulting gravitational force on the central parts of the sphere is therefore zero, and they remain stationary. At a certain distance from the center, however, a mass is subjected to asymmetrical interaction of gravitational forces, resulting in a force aimed at the center. It can be mathematically proved that the force thus generated is proportional to the distance from the center. (This law is applicable as far as the surface of the sphere; beyond that the force diminishes in inverse proportion to the square of the distance from the sphere's center.)

Under the influence of gravitation every part of the sphere moves toward the center. Since the force, as already mentioned, is proportional to the distance from the center, the speed of this inward motion is proportional to the same distance. How would the motion appear if we were to observe it from a part of the sphere far removed from the center? A mathematical computation would readily show that this motion looks the same everywhere: relative motion is directed toward the observer and speed is proportional to the distance.

This law of motion recalls Hubble's law, which states that the galaxies are receding at speeds proportional to their distances from us. The only difference is that in our model the motion is *toward* and not *from* the observer. The sphere of ambiplasma within our focus thus contracts in obedience to the same law our metagalaxy obeys when it expands. The law of contraction may be derived without sacrificing the assumption we have introduced: in a sphere originally in a state of rest, where mass is uniformly distributed, the general law of gravity is operative.

Since every part of the sphere falls toward the center at a speed proportional to its distance, it obviously follows that all parts will reach the center at the same time. The whole mass of the sphere would supposedly be clustered at that one spot. But, as we already noted in Chapter II, this is an unrealistic conclusion. Having chosen to adopt a highly idealized model, we must be prepared to pay the consequences. We assumed an exactly uniform distribution of mass and started out from a state of absolute rest. Should we elect to adopt a more realistic model, it will necessarily be more complicated as well. For instance, we could assume an uneven distribution of the mass and some stray motion even in the primordial state. Even though deviations from the idealized model are minor at the outset of contraction, the difference grows automatically by comparison with the more realistic model; the closer we get to the moment when the whole mass supposedly clusters at one spot, the greater the deviations. Yet long before this moment is reached, our idealized model has ceased to relate to what could be a physical reality.

In Chapter II we saw that these deviations are of essential importance in a model where the metagalaxy consists of koinomatter only. In the ambiplasma model we are considering here, deviations from the ideal *in this respect* are of less importance because the motion is changed by new forces, unleashed by annihilation, long before we reach the single-spot state. The contraction attainable in our *idealized model* never reaches such a state as to make the model unrealistic on that account. We can still use the ideal model, at least for the time being. Later we shall be discussing other phenomena that limit the model's usefulness.

To make our model more specific, we suggested conceivable values for the radius (a trillion light-years) and density (one particle in a cube with sides of a hundred meters). The magnitude of these figures may be taken to typify an early state of development. It need not be the primordial state, but rather the state at a point of time when the model becomes relevant for us. We beg leave to sidestep the question, "What happened before then?"

Owing to the extremely low density, the gravitational force is extremely small. The already described motion of falling toward the center of the sphere begins slowly and imperceptibly. But as billions of years pass in a near-empty and apparently uneventful world, the inward motion picks up speed and the contraction proceeds apace. When the resulting density has increased to (say) one particle per cubic meter, some chance of collision between a proton and an antiproton becomes evident. The radiation from annihilation becomes significant (see Figure 13b). Every part of the enormous plasma mass produces gamma rays and, in consequence of magnetization, radio waves as well (see Chapter V). This radiation is absorbed and reflected within the ambiplasma as it seeks its way toward the surface of the sphere.

When radiation impinges on a body that reflects or absorbs it, it exerts pressure on the body. This "radiation pressure" can be variously interpreted. Since gamma and light rays—and radio waves, too—consist of rapidly moving photons, the pressure may

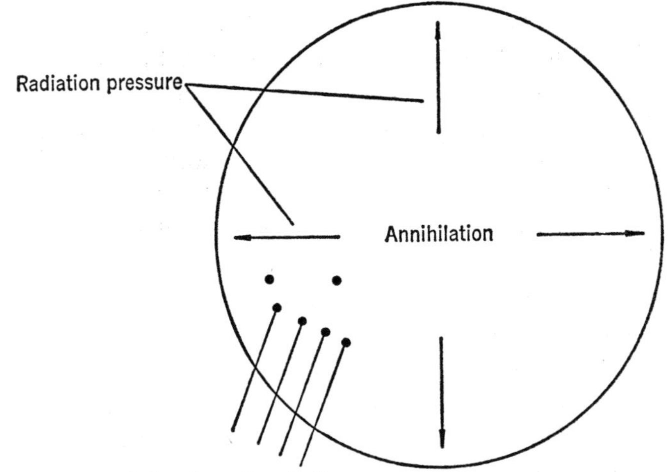

Figure 13b. A galactic model. (2) When the sphere has been compressed sufficiently, the annihilation becomes a significant factor. This produces an outward-directed radiation pressure, which halts the inner motion and converts it into an outward motion.

be regarded as a result of collisions between the photons and the body's atomic particles. If one volume, *A*, emits radiation which is absorbed by another volume, *B*, and *B* returns just as much radiation, which is absorbed by *A*, the cross fire of photons generates a mutual repulsion of *A* and *B*. The force of repulsion is inversely proportional to the square of the distance: in other words, it obeys the same law as gravitation between *A* and *B* except that radiation pressure repels and gravitation attracts. Accordingly, radiation pressure has the effect of reducing gravitation. If the pressure is strong enough, it will neutralize gravitation entirely or even overcome gravitation and force *A* and *B* to repel one another.

Therefore, as soon as annihilation becomes a significant factor in our ambiplasma sphere, the gravitational force will be apparently reduced. The increase in speed will then proceed at a slower rate. When the density has increased even more, the radiation from annihilation becomes so intense as to neutralize the effects of gravitation altogether. From that moment the speed of contraction begins to decrease, but by then it will already have reached a very high value. With continuing contraction the density increases, and with it the radiation from annihilation, at an ever faster pace. Our world, which for a trillion years was so uneventful, has entered a dramatic phase. This phase will last for billions of more years on the cosmological clock.

Radiation pressure now increases at an ever faster rate and its force will be many times that of gravitation. Before long it will be strong enough to check the contraction and reverse the motion from inward to outward. Our metagalactic plasma sphere has reached its minimum size, and from now on will expand at an accelerated pace (see Figure 13c).

It is easy to demonstrate theoretically that the outward speed must obey a law similar to the previous inward speed: it is proportional to distance from the center, but the constant of proportionality is not necessarily the same as before. Our

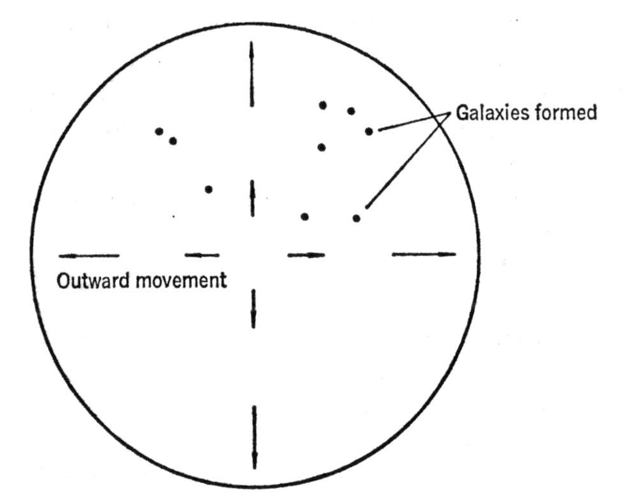

Figure 13c. A galactic model. (3) The metagalaxy reaches its present state, which is characterized by an expansion, observed as a red shift.

theoretical result agrees with Hubble's law (which of course is based on observations).

As the ambiplasma sphere expands, its density lessens. This tendency is reinforced by annihilation of much of the ambiplasma, indicating that mass has been transformed into radiation. A large part of the energy thus released goes toward producing the outward motion. The lowered density reduces the annihilation and, when the sphere has expanded far enough, the outward radiation pressure becomes so small that gravitation again gets the upper hand. The outward speed, instead of increasing, starts to decrease in consequence of gravitation.

In theory the final stage may be of two types. If the outward speed caused by radiation pressure is great enough, the inward force of gravitation cannot arrest it and the expansion continues on toward infinity. A final phase of this type produces a sphere which expands indefinitely; density falls off, and annihilation is again on a very small scale. The dramatic event fades into memory; no repeat performance is ever given. The violent "radiation explosion" that caused the reversal has been accompanied by a loss of mass. About half the mass (more or less, depending on the circumstances) has been annihilated. (As we shall see later, it is this type of process which the metagalaxy has experienced.) The other type of final stage occurs if radiation pressure does not impart sufficient outward speed to the ambiplasma, so that gravitation will slow down its outward motion and eventually reverse the direction. In that case the same sequence of events is repeated. A new contraction takes place and is followed by a new radiation explosion, which checks the inward motion and sets the stage for a second expansion. This expansion is again slowed down by gravitation, and a third process of the same type begins. Thus, the sphere is pulsating between a minimum and maximum size. But every time it reaches the minimum, a violent annihilation process ensues. A part of the mass is thereby consumed, and the mass of the sphere will accordingly diminish with each pulsation. The end product of repeated pulsations will be complete annihilation of the mass. However, the second, pulsating alternative is probably not the one that corresponds to the real world.

THE MODEL AND THE METAGALAXY

We have tackled a complex combination of problems with a method common in physics: construct a model so simplified as to permit exact mathematical analysis, yet at the same time related as closely as possible to physical reality.

Ours has been a very simple model, based on the principle of symmetry between koinomatter and antimatter. We have studied how a sphere filled with a magnetized, uniform ambiplasma develops under the influence of gravitation. We found that the sphere first contracts and then, under the influence of the radiation pressure generated by annihilation, begins to expand. Theoretical calculations show that the expansion must obey the very law derived by Hubble from his observations of the galactic red shift. In that respect the model satisfactorily describes an essential property of the metagalaxy.

However, our next comparison between model and metagalaxy indicates that the model must be further developed in at least two respects to make it agree with the metagalaxy's most important properties. The first property is nonuniformity. We assumed that initially the density is uniform throughout the volume, and we shall not alter this assumption. But in dealing with development of the ambiplasma, we also premised no change in the uniform density. This assumption, obviously, gives us too crude a picture of the metagalaxy. Actually, it consists of 10^{10} galaxies, with densities enormously greater than the space between them. We thus disregarded the fact that the metagalaxy presents a "granulated" pattern, with each granule representing one galaxy. To approximate reality more closely, we must therefore study how the galaxies are formed. Second, in our model we assumed that koinomatter and antimatter are in perpetual blend. This is a very reasonable assumption to make for the primordial state, but it just won't do for the present state of the world. Our sun and solar system contain koinomatter exclusively. As for the stars, we cannot say that any one star consists of koinomatter or antimatter, but we do know there is no room for both in the same star. If our model is to bear any relation to the present-day world, we must allow for processes that separate koinomatter from antimatter. These processes, moreover, must be operative while the metagalaxy is developing.

THE FORMATION OF GALAXIES

Earlier we discussed the conflict that occurs whenever a problem in physics is subjected to theory. The construction of a model that is supposed to represent physical reality always has to take two conflicting pressures into account: on the one hand, it must be simple enough to make it amenable to mathematically stringent methods; on the other hand, it must be sufficiently realistic that the results of studying the model's properties actually lend themselves to the problem at hand. The simplicity of the model—its mathematical beauty—is incompatible with its faithfulness to reality. The beautiful are seldom faithful and the faithful are seldom beautiful.

We ran into this conflict in our discussion of the big-bang theory. The question of how ylem could be formed was found to presuppose such an extremely idealized model that serious doubt could be cast on its ability to reproduce the essential elements of a physical reality. By contrast, the ambiplasma model we introduced earlier in this chapter escaped the same difficulty because it never achieves the enormous concentration of the big-bang model. Our model, however, has other shortcomings. A major shortcoming is its assumption of complete uniformity; actually, the metagalaxy is granulated, or divided into galaxies.

It has long been known that a mass of gas (in the new terminology, a plasma) which uniformly fills a volume of cosmic dimensions is in an unstable state. Even though the original distribution of the plasma mass may have been highly uniform, small variations in density must inevitably develop. A region of greater density will then build up a gravitational field, causing the nearby plasma to fall inward. As a result, density is increased in a region which was already more dense than its surroundings.

If this process occurs everywhere in the originally uniform medium, a pattern of "granulation" is formed. The cloud mass splits up into a number of local condensations. However, the process is counteracted by other phenomena, including thermal motion and magnetic fields. A study of these problems must weigh the concentrating tendency of gravitation against the effects working in an opposite direction.

Problems of this kind first arose in attempts to explain the origin of the stars. It was thought—and still is, for that matter—that the sun and nearby stars were formed from a cloud that constituted one part, perhaps a relatively small part, of our galaxy, say on the order of 100 or 1,000 light-years across. Although we have yet to arrive at a completely clear picture of how stars are born, the earlier stages of this process are better understood. Thus we know fairly well how an originally homogeneous mass of gas becomes unstable and begins to concentrate. It is in trying to explain the later stages that the really serious difficulties arise.

By logical extension, the formation of galaxies should occur in observance of the same general principles that hold for star formation. Our point of departure, therefore, is a uniformly distributed plasma, but of enormously greater dimensions, on the order of one billion light-years. In the professional jargon, this mass becomes "unstable toward gravitational disturbances." The resulting contraction into clumps probably signifies the beginning of galactic formation, a process whose main features are comprehensible because we have already studied a similar process that produces stars.

But even this approach to an explanation eventually leads us into serious difficulties. To begin with, the analogy with star formation is of little help because our grasp of its later phases is still dim. Further, we should not expect any major resemblance because the end product, a galaxy, differs so much from a star, and not only in size. Even more serious is that the theory of star formation assumes that the condensing mass consists exclusively of koinomatter. The theory, of course, readily lends itself to antimatter, but it falls down when confronted with a mixture of koinomatter and antimatter: an ambiplasma. By its very nature, ambiplasma must incur annihilation, which may be of fundamental importance.

We noted earlier that annihilation plays a negligible role as long as the density of ambiplasma is very low, but, as density increases, so does annihilation. Since it is proportional to the square of the density, annihilation accelerates rapidly with increasing density. The theory of star formation should therefore be applicable only to the first stage of galaxy formation, when ambiplasma density is still very low. What we can say is that we fairly well understand how the originally uniform ambiplasma in our model is bound to become granulated—that is, divide into a large number of parts that gradually evolve into galaxies. But the further development of the galaxies poses a much more formidable problem.

It is hard to pinpoint that stage of metagalactic development when galaxies start to form (see Figure 14). The first tendencies toward clustering may well occur at a very early stage, and galaxies may be forming while the metagalaxy is contracting. We thus have two processes occurring simultaneously, and the rate at which they develop will require careful study before we can speak with any certainty about the over-all course of events. Besides, the two processes are very much interwoven the whole time. Contraction of the metagalaxy alters the premises for the local condensations that give rise to the galaxies. Conversely,

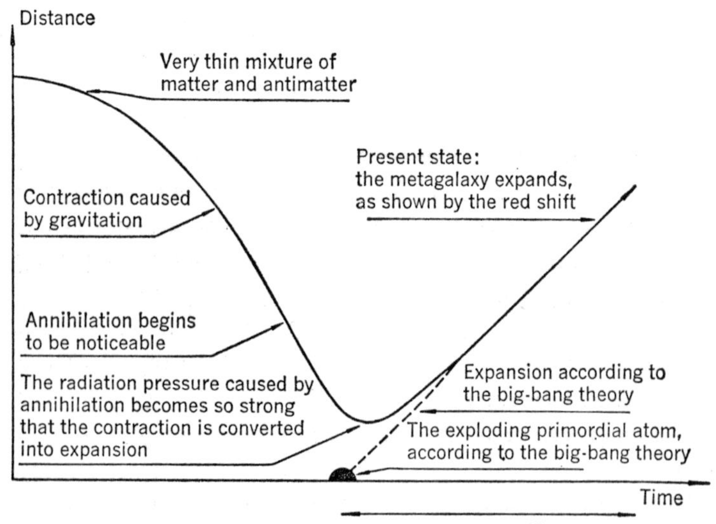

Figure 14. The development of the metagalaxy.

the formation of galaxies alters the general development of the metagalaxy, since the formation entails increased annihilation and hence radiation.

Our simplified metagalactic model assumes that ambiplasma density is uniform throughout the volume at any given time. Annihilation then becomes no more than a function of the size of the metagalaxy. When maximum density is reached with minimum size, the annihilation is also at maximum; when the metagalaxy again expands, annihilation quickly declines to a near-negligible value. But the assumption fails to hold as soon as galaxies are formed, since their density greatly exceeds the average density and the density in intergalactic space is much less. It follows that annihilation will increase within the galaxies and decrease between them. However, the increase within the galaxies outweighs the decrease, so that the end result is an increase in annihilation. A discussion of the following extreme case will show why this must be so.

Suppose that all the ambiplasma is concentrated in the galaxies, whose combined volume is 0.001 that of the metagalaxy as a whole. The intervening space is then completely empty. For the sake of simplicity, we take one galaxy and assume it to be uniform: its density will then exceed the mean metagalactic density by 1,000 times. Since the annihilation per unit of volume is proportional to the square of the density, it will increase one million times. But as the total volume of the galaxies is only one-thousandth that of the metagalaxy, the total annihilation comes to 1,000 times (0.001×10^6) what we would get in the absence of galaxy formation. We thus see that the annihilation radiation and hence radiation pressure can increase very sharply by virtue of such galaxy formation. This process, therefore, must be of great importance for development of the metagalaxy as a whole.

The above explanation does not invalidate the theory of pulsation, of expansion followed by contraction and then by expansion again. Neither does it jeopardize the theoretical derivation of

Hubble's law. But it does mean that maximum radiation may well take place after the metagalaxy changes course. It can also mean that more total mass is lost than in the uniform model. This is a very important question.

If, under the influence of gravitation, the metagalaxy could contract without interference from radiation pressure, the whole mass would be annihilated. What saves the metagalaxy is the radiation explosion. But the death sentence is barely commuted when a new peril appears. Owing to the formation of galaxies, local concentrations of ambiplasma are formed. As its density is thereby increased, the annihilation picks up speed. The once so evenly shining metagalaxy breaks up into a number of luminous spots (each "spot" understood here to mean a region the size of a galaxy), which in addition to light emit gamma rays and, even more, radio waves. If gravitation is the only force of importance in the formation of galaxies, the ambiplasma must be increasingly concentrated until it is almost completely destroyed. Unlike the metagalaxy, the galaxies cannot be saved from this fate by an expansion caused by radiation pressure. They are too small for such a process to be effective. (If we apply the case of pulsation discussed earlier in this chapter, it leads to their ultimate doom.)

THE SEPARATION OF AMBIPLASMA

The only process that can save a galaxy-forming ambiplasma from total annihilation is a *separation* of its koinomatter and antimatter. If a region of ambiplasma is divided into one area containing only koinomatter and another area containing only antimatter, annihilation will virtually cease. This is because the two areas can be kept apart by a thin Leidenfrost layer, in which only very weak annihilation takes place (see Chapter V). The galaxies may therefore be saved from self-destruction if their

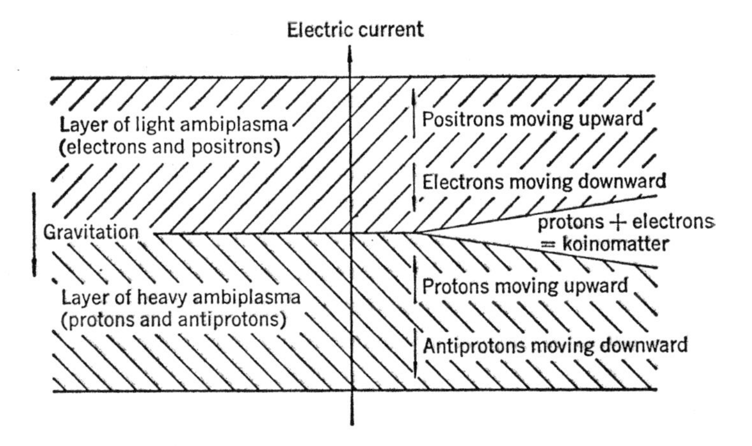

Figure 15. The separation of matter and antimatter. If, under the action of gravity, a layer of heavy ambiplasma is situated at the bottom of an atmosphere, the top of which consists of light ambiplasma, an electric current produces a separation of matter and antimatter. In the light ambiplasma the current pulls electrons toward the interface, and in the heavy ambiplasma the current brings protons to the interface. Consequently, at the interface koinomatter (= protons + electrons) is accumulated.

formation is accompanied by a process of separation (see Figure 15).

As noted earlier, such a process is also required to explain the evolution of the metagalaxy into its present state, which at least in certain regions, including our own, is characterized by the existence of koinomatter unmixed with antimatter.

The process we are looking for need not lead to a complete separation of the two. Suppose that we have a volume with 100 units of ambiplasma, composed equally of koinomatter and antimatter. Let us further suppose that a process operates which concentrates koinomatter in one half of the volume, *A*, and antimatter in the other half, *B*, so that *A* comes to contain 80% of koinomatter (or 40 units) and 20% antimatter (10 units); the proportion is reversed in *B*. Annihilation will continue until the 10 units of antimatter in *A* have destroyed 10 units of the koinomatter, leaving 30 units of the latter; in *B* the annihilation leaves 30 units of unmixed antimatter. All in all, 40 units have been annihilated.

In the formation of galaxies, the ambiplasma faces the possibilities of being separated or being destroyed. If we can find a process that likewise concentrates even a part of the koinomatter in certain areas, we shall have saved a part of the plasma from extinction. When the annihilation is done and finished, there will be left areas of pure koinomatter and pure antimatter. We thereby arrive at a state of the universe which could resemble the one we have now.

We must now ask: What physical effects are capable of producing a separation of the type we are looking for? Given certain general conditions, separation may be shown to ensue if a magnetized plasma is located in a gravitational field. Several more or less allied problems are involved here, but a systematic study of them has not yet been made. It is therefore too early to attempt a more detailed model of the process or to try to pinpoint where and when it takes place. Here we shall confine ourselves to a simple model in order to demonstrate the type of phenomena that can produce the all-important separation.

AMBIPLASMA IN GRAVITATIONAL FIELDS

In the atmosphere above our earth, air pressure diminishes with increasing altitude. It falls to half the sea-level pressure at an altitude of 5 km and to one-fourth at 10 km. The closer to earth, the more compressed the air becomes because of gravitation This effect is compensated by the thermal motion of the air molecules. In the absence of gravity the molecules would fly straight out into surrounding space. In its normal state, therefore, the earth's atmosphere represents a compromise between gravity, which tends to pull all molecules down to the surface, and thermal motion, which would scatter them out into space.

The atmosphere consists chiefly of nitrogen, with an admixture of 20% oxygen and a few other gases. If we were to substitute hydrogen, whose molecules weigh only 1/14 as much as the nitrogen molecules, the compressing effect of gravity would lessen. Other things being equal, the pressure in a hydrogen atmosphere would not be halved until we came to an altitude of 70 km. If an "ideal atmosphere" (by which we mean one without solar radiation, winds, and other "disturbances") consisted of blended hydrogen and nitrogen, the two gases would be independent of one another in the sense that the concentration of nitrogen would diminish to half at 5 km and that of hydrogen at 70 km. This means that gravity would concentrate the nitrogen close to the earth's surface but would permit the lighter gas, hydrogen, to disperse more freely. If the two gases were present in equal amounts, most of the nitrogen would hover close to earth, with most of the hydrogen higher. Generalizing the picture, we could say that our atmosphere had a layer of nitrogen superimposed by a layer of hydrogen.

Let us now shift focus to an ambiplasma under the influence of gravitation. Obviously, the earth's gravity serves little use to us now; we must think instead of a gravitational field in a galaxy or in some cosmic cloud (we will leave the latter's identity unspecified for the time being). The ambiplasma is a mixture of light gas, consisting of electrons and positrons, and a gas 1,840 times heavier, consisting of protons and antiprotons. Note that the difference in weights is much greater than that between hydrogen and nitrogen. Gravitation will therefore greatly concentrate the proton-antiproton plasma (we shall call it the *heavy ambiplasma*) but has only a negligible effect on the *light ambiplasma* (electrons and positrons). To draw a generalized picture once again, we can say that we get a layer of heavy ambiplasma at the bottom (toward the source of gravitation) and a layer of light ambiplasma on top of it.

What happens in this mixed ambiplasma if we insert a *vertical* electric current in a direction opposite to the gravitational force? A current in a plasma makes the charged particles move—the positive in the direction of flow, the negative in the opposite direction. An "upward" current in the light ambiplasma induces upward motion in the positrons and downward motion in the electrons, with the boundary between light and heavy ambiplasma as their target. Meanwhile, in the heavy ambiplasma, the current moves the negative antiprotons downward, while the positive protons head upward toward the boundary with light ambiplasma. This border area is a meeting place for the electrons from above and the protons from below; in other words, the two koinomatter particles are collecting there. At the same time the antimatter, or the positrons and antiprotons, are being far removed from this rendezvous. At the boundary, therefore, we get an area where the koinomatter of ambiplasma is concentrating. The process is highly suggestive of electrolysis, in which the conduction of an electric charge through a solution liberates certain substances in the solution.

If the current is in the opposite or downward direction, the boundary will be a collecting place for antimatter rather than koinomatter.

We thus find that a vertical electric current in an ambiplasma can induce "electrolytic" separation of koinomatter from antimatter in a gravitational field. The existence of an electric current does not introduce a new assumption, but is a consequence of our earlier assumption that the ambiplasma is magnetized. The motions of a magnetized ambiplasma must, as a matter of course, induce electric currents. Indeed, we could say that, by assuming magnetized plasma, we have also assumed that electric currents flow in it. Magnetic fields and electric currents are so intimately associated that it is impossible to think of one without the other. Since, generally speaking, an electric current should have a vertical component (ignoring only the special case where current flows in an exactly horizontal direction), we can account for a separation taking place under very general conditions.

The process described is far from being the only one which can achieve separation. For instance, there need be no direct dependence on an electric current. If a plasma in a gravitational field has a magnetic field which shows a certain heterogeneous pattern, the electrons and protons may both move in the same direction, while the positrons and antiprotons take off in the opposite direction. This means that the koinomatter and antimatter in an ambiplasma are separated by "gliding apart."

When different processes of these types are studied, it emerges that they ought to be extremely effective on a small scale—small, that is, from the cosmological point of view. But it is quite a different thing to find a process that can separate enough ambiplasma to give us a whole galaxy of pure koinomatter or pure antimatter. One stumbling block is that separation on a large scale demands transportation of koinomatter particles over huge distances away from the particles of antimatter. Considering the time and the forces available, it is unlikely that the transporting mechanisms could cope with such a task. This does not rule out the possibility of separation on a galactic scale. It could conceivably start off modestly, giving rise to many small areas of separated koinomatter and antimatter; later, all the small koinomatter clouds combine, and a corresponding process combines the antimatter into another big cloud. However, all these processes have still to be analyzed in depth; until then, our discussion cannot be more than loosely speculative.

In Chapter V we found that the requirement of symmetry between koinomatter and antimatter could be satisfied by having every second galaxy consist of antimatter or by having every second star in our galaxy (and in other galaxies) consist of antimatter. Obviously, the existence of antimatter in remote galaxies is less open to emotional objection; its existence in stars closer home would really give us something to worry about. Our analysis of the separation process, however, results in a different order of priority: it is much more difficult to see how a whole galaxy could be separated as compared with small regions within a galaxy.

We have discussed possible separation processes only in very general and rather vague terms. That is because they have not been analyzed in greater detail. The time is therefore not ripe for more specific models. It is likely, however, that separation is closely related to the formation of galaxies. Our lack of knowledge in this field is one of the weakest points in our whole discussion of antimatter in the cosmos.

VII

The Cosmological Problem

COSMOLOGICAL SYNTHESIS

In looking back upon the progress of science, we are struck by the interwoven histories of astronomy and physics: time and again, the course of astronomy has inspired physics, and, just as often, the inspiration has arisen in some development of physics. Every attempt to determine the temperature or chemical composition of the stars was doomed to failure before spectroscopy came along. Until nuclear physics made us aware of the tremendous energies unleashed by nuclear reactions, every attempt to understand how the stars produced their energy was likewise futile. Who knows but that the discovery of the positron and antiproton, and the irresistible tendency of elementary particles toward symmetry, may hold the keys for revealing the truths of cosmology?

We started with the assumption that symmetry governs the distribution of koinomatter and antimatter in the cosmos. The assumption is not a necessary consequence of elementary-particle physics, but it is a reasonable one to make. Rather than introduce new natural laws in our analyses, we have tried to find out how much we can understand with the help of laws already known. Hence our efforts to incorporate cosmology in the idea structure

Table 4. Fields Common to Physics and Astronomy

Physics	*Common Fields*	*Astronomy*
Pendulums Falling motion	Mechanics	Orbits of moon and planets
Atomic structure	Atomic physics Spectroscopy	Spectroscopic methods used to determine the chemical composition, the temperature, and the motion of celestial bodies
Electric currents and magnetic fields Electric discharges in gases	Electricity Plasma physics	Sunspots, solar activity, aurorae, cosmic radiation
Structure of atomic nucleus Atomic energy	Nuclear physics	Generation of energy by the sun and stars
Properties of elementary particles	Elementary particle physics	Cosmology based on matter-antimatter symmetry

of laboratory physics. We have also foregone attempts to explain how the universe began and where its limits are—assuming that there are any limits. We have modestly confined ourselves to the latest trillion years of time and the nearest trillion light-years in space.

We began with what we called a primordial state, characterized by a rarefied ambiplasma in that part of the universe under our observation. The ambiplasma, we pointed out, was under the influence of gravitation. It is not necessary to ask how this state arose. If importuned to answer this question, we could refer back to an even more primordial state. As was observed in the first section of Chapter VI, we must then assume that only radiant energy was found in space. This radiation generated the proton-antiproton pairs by a mechanism described in Chapter III. We could thus account for the origin of ambiplasma.

But regardless of how far we might pursue questioning on this point, our term of reference is still a large volume of ambiplasma, which we intend to follow and see how it develops into the present metagalaxy. For the sake of simplicity, our model shall be a sphere. It contracts until density reaches a point where annihilation becomes an important factor. The annihilation generates radiation, particularly in the form of gamma rays and radio waves. The pressure imparted by this "radiation explosion" stops the contraction and changes it to expansion, which is occurring at the present time, as is manifested in the red shift of the galaxies.

At the same time a condensation process is occurring within the metagalaxy (by virtue of gravitational irregularities), whereby the galaxies are formed. At the beginning of their existence the galaxies consist of ambiplasma. When this contracts it gives rise to intensive radiation. The separation of koinomatter and antimatter saves the galaxies from total destruction. When that part of the ambiplasma that is not separated begins to burn itself out, the emission of gamma rays and radio waves from the newly formed galaxy decreases. In due course the galaxies develop into the "normal" state they have at the present time.

A physicomathematical calculation of the properties possessed by our simplified metagalactic model makes it possible to compare them with the present state of the metagalaxy. Data from nuclear physics tell us how great a density ambiplasma must have to produce an annihilation forceful enough to reverse the process of contraction to expansion. It appears that the density at the turning point of the metagalaxy must be of the order of 1000 particles per cubic meter. This is about 10,000 times the present mean density (roughly arrived at by dividing the total mass of the galaxies by the volume of the metagalaxy). We infer that the metagalaxy now has a volume 10,000 times greater than when it was of minimum size. Accordingly, the metagalactic diameter has increased $\sqrt[3]{10{,}000}$ or 22 times, which also means that the mean distance between the galaxies has increased by about the same factor.

In those respects that admit of quantitative comparison between theory and observations, the correlation is fairly good. We must bear in mind, however, that the theoretical findings are derived from a highly simplified model; not only that, but the observations are necessarily very indeterminate in certain respects. Quantitative proof of the theory will require a great deal of work by theoreticians and observers.

The formation of galaxies is crucial to metagalactic development, and it would be beyond the scope of this book to elaborate upon it. But given our starting points, it is likely that a contraction of ambiplasma is the first step in the formation of galaxies. The embryonic phase should therefore be marked by intensive annihilation, together with the emission of gamma rays and radio waves. In addition, of course, the whole galaxy will generate tremendous heat, giving off highly luminous light. As noted earlier, our measuring instruments are largely insensitive to gamma rays. The initial state of a galaxy's development therefore has to make itself known to us by means of intensive radio and light emissions. In the past few years astronomers have discovered remote high-temperature radio stars, which at the same time emit very strong light. These are the *quasars*. The amounts of energy coming from them are so large that the only reasonable explanation for their source is annihilation. It is therefore quite possible that quasars are galaxies in the act of being born. (The curious reader will find almost as many theories about quasars as there are articles written about them!)

As the separation of koinomatter and antimatter proceeds and part of the nonseparated ambiplasma is annihilated, the galaxies enter a calmer stage. It is conceivable that this process relates to an ordinary galaxy which emits radio waves. The radio emission, it turns out, often does not emanate from the galaxy itself, but from two sources on either side of it. There may indeed be only one ultimate source, the light ambiplasma (electron-positron gas), which escapes from the galaxy when gravitation is in no position to stop it. That it streams forth in two diametrically opposed directions suggests the effect of electromagnetic forces.

When most of the ambiplasma is annihilated, the galaxy comes to its third stage, the one we have at present. Normally, the emission of radio waves is fairly negligible, but sources of enormous energy are available if koinomatter and antimatter should happen to intermix. And there are many energy-generating celestial phenomena whose cause is a mystery to us. To take just one example, we still lack a satisfactory theory for supernovas.

THE THEORY OF RELATIVITY

When Einstein's theories caught hold at the beginning of this century, they revolutionized many old conceptions. The cosmological consequences of the general theory of relativity stirred the scientific imagination tremendously. They stimulated discussion of universal structure and touched off many intriguing speculations.

The first of Einstein's theories, the special theory of relativity, has already become a classic legacy of physics. Now that the laboratories so often work with accelerators, where the speeds imparted to electrons and atomic nuclei approach the speed of light, the special theory of relativity is in everyday use. The theory states that no particle of matter can actually attain the speed of light, and it minutely describes the behavior of the fast particles in our accelerators. The equivalence of mass and energy (on which we drew for our discourse on annihilation) and the radio waves emitted by spiraling electrons and positrons are both theoretical consequences of the special theory.

The second or general theory of relativity was primarily developed to explain gravitation. However, the observable differences between Einstein and Newton on the same subject are very small (a slight discrepancy in the revolution of Mercury around the sun, the extra deflection of a beam of light passing near the sun, a red shift of the spectral lines in a gravitational field). For this reason the general theory has not been applied to celestial mechanics on an appreciable scale. The simpler Newtonian theory is still employed almost exclusively to calculate the motions of celestial bodies. Also, the general theory of relativity is too complicated to be used for calculating the motions of electrically charged particles. Accordingly, since plasma physics and magnetohydrodynamics have become more important in the field of cosmic physics, the general theory has become less serviceable.

However, the general theory of relativity has played a major role in cosmology because of the new vistas it has opened concerning the structure of the universe. The mathematics of this theory introduces a "curved space," which leads to analogy with a curved surface. The difference between ordinary or Euclidean space and curved space resembles the difference between a straight and curved line, as on the surface of the earth. If the earth were as flat as a table, we would eventually come to an edge if we traveled far enough in any one direction—unless the earth were of infinite size. But since the earth's surface is curved, it need not be infinite even though we never come to an edge. Were we to continue traveling "straight ahead"—that is, without turning either to right or left—we would come back to where we started after a trip around the world.

Reasoning by analogy, we could postulate a space of finite extent, yet without limits. If we were to travel "straight ahead" we should eventually come back to our starting point. Although no "limit" would be encountered in the journey, we could not claim that space was infinite. (In the expanding models of the universe which are of interest to us, however, the journey would require a length of time greater than the duration of the expansion.)

This image of the universe, so fascinating in many respects, is *possible* according to the theory of relativity, but it is not an inevitable consequence of the theory.

Even if the general theory of relativity is correct, which we have no reason to doubt, space may very well be infinite. The finite-versus-infinite argument cannot be resolved in the armchair, but only in the astronomical observatories.

The *direct* observation of space for this purpose is impracticable, but two alternatives utilize certain different combinations of observable data (such as the Hubble constant and mean density in space). If we attach the most probable values to these in the theoretical formulas, our conclusion is that *space is infinite.* However, this result is valid only for that part of the universe having the size of the metagalaxy. We can state that our metagalaxy cannot be the whole of the universe, but there is still a possibility that we have a finite world; however, it would have to be much larger than the metagalaxy. In terms of the problems of metagalactic structure, which concern us here, we can accordingly assume that space is infinite.

In describing the development of the metagalaxy, we did not draw directly on relativity theory. The essential developmental features can be described without it, and it seemed better to neglect relativity, to keep our presentation as uncomplicated as possible. (In so doing, we deliberately turned our back on the romantic mystery in which many love to enshroud relativity theory.) It goes without saying that the theory of relativity must play a very important role in any detailed cosmology. The energy released in annihilation is an obvious example of the equivalence between matter and energy postulated by the special theory of relativity. Relativistic effects also operate in limiting the size of the mass that forms the metagalaxy: since in the first stage of metagalactic development the speed of contraction is proportional to the distance from the center, there would be no limit to the velocity if the original plasma sphere were very large. But here certain relativity effects become important, putting an upper limit on the size of the metagalaxy.

The relativistic effects are most important when the metagalaxy changes from contraction to expansion. Space becomes highly "curved" and to some extent the surrounding space loses contact with the metagalaxy. However, it would take us too far afield to discuss these rather complex phenomena.

There is no need for us to get bogged down in a discussion of a four-dimensional space-time continuum. The main features of metagalactic development can be described very well in terms of ordinary three-dimensional space and of ordinary time.

CHARLIER'S MODEL OF THE UNIVERSE

In 1908, C. V. L. Charlier suggested that the universe is composed of systems of increasing dimensions. The stars form clusters, the clusters form galaxies, the galaxies form groups, and the groups form a metagalaxy (the last two stages in this sequence were unknown in Charlier's time). According to Charlier, the systems would continue to grow larger and larger unto infinity.

One of the sources of this model was the "Olbers' paradox." When astronomers began to explore the fixed stars, they found that those in the sun's vicinity were more or less uniformly distributed. From this they concluded that the uniform distribution of stars held for the universe in general. But if the universe is infinite, the conclusion leads us to an absurdity. Let us assume a certain quantity of light here on earth from the stars which are between 10 and 20 light-years away. Then double the distance and consider the stars which are between 20 and 40 light-years from us. If the stars are uniformly distributed in space, the number of stars in the 20 to 40 range would outnumber those in the 10 to 20 range eight times. But because the former stars are twice as far away, each gives us only one-quarter the light, since luminosity decreases with the square of the distance. The result: we get one-fourth as much light from each of eight times as many stars; in other words, the stars between 20 and 40 light-years away give us twice the light as the stars between 10 and 20 light-years away. When we proceed even farther out in space, we similarly perceive ever more light from an increasing number of remote stars.

As Olbers pointed out, the remote stars in an infinite space, assuming their uniform distribution, would emit light of such intensity that the night heavens would shine like the sun.

We have already noted that the stars are not uniformly distributed in space. When we come to a distance of several thousand light-years from the sun, we reach the limits of our galaxy (which is 10,000 light-years in thickness and 100,000 light-years in diameter). Beyond that there is intergalactic space, where the stars are so few in number that there is no danger of their blinding us with their light.

However, the problem repeats itself when we consider the other galaxies. To be sure, the nearest one, in Andromeda, is barely visible to the naked eye, and light from the others is even feebler. But if the galaxies were evenly distributed over an infinite space, what then? Wouldn't the combined intensity of their light be tremendous? Once again we escape this prospect by virtue of the fact that the metagalaxy is limited (besides, the red shift operates to diminish light intensity).

If Charlier's idea were applied to what we have arrived at here, the same problem could be repeated over and over again. There might be many other metagalaxies beyond our own, which together form an even larger system: a "teragalaxy," to coin a new word (from the Greek *teras*, marvel or monster; compare Table 1). A large number of teragalaxies would then comprise a bigger system, and so on ad infinitum. If certain of Charlier's mathematical criteria are met, a similar progression of systems gives a total light intensity which stays within reasonable bounds. The Olbers' paradox would be avoided, as well as one other difficulty—keeping gravitation from the remote regions from becoming too great.

Charlier's universe is thus infinite. The total mass of the universe is also infinite, but its mean density has a very low value (mathematically, we can say it approximates the value of zero). Most of the night light we receive comes from our local system, from the stars (apart from the sun) in our galaxy.

Little attention was first given to Charlier's idea because it happened to be presented at the same time as the general theory of relativity, with its fascinating suggestion that the universe might be finite yet unlimited. Since observational findings no longer support so convincingly the finite universe of relativity theory, the Charlier model takes on special interest.

The metagalactic theory we have been discussing is in line with Charlier's model. Presumably, however, Charlier thought that the various systems were static, whereas our model of the metagalactic system is dynamic. Our theory, of course, describes how the system is formed, contracts, and expands again.

But having said this, we should also point out that our theory of metagalactic development can be reconciled with a finite universe of the type posed by relativity theory. Such a universe, however, would have to be much larger than the metagalaxy.

THE ORIGIN OF THE ELEMENTS

There are 91 natural elements on earth, and about a dozen "artificial" elements can be produced in accelerators. Spectroscopic analyses have disclosed that the sun and stars are structured of the same elements familiar to us on earth. Some of them, such as hydrogen and helium, are much more common in the sun and stars. Hydrogen is the most common element in the observable universe, comprising more than 90% of all the atoms. The next most common element is helium. As we noted earlier, spectroscopic analysis of a star does not permit us to say whether its elements are of koinomatter or antimatter.

If we want to make the simplest possible assumption about the primordial state, we should assume that space contained an ambiplasma, which in addition to electrons and positrons included an equal number of protons and antiprotons, but no heavier elements. We might take this state further back in time to something even more primordial, when space contained only the radiation that produced protons and antiprotons in pairs. The assumption must be confined to protons and antiprotons, since radiation is incapable of producing heavier atomic nuclei directly.

If the original ambiplasma had developed into the present state of the universe by means of the processes we have studied, we should first of all expect our world to consist exclusively of hydrogen gas. The expectation is more than 90% correct, inasmuch as more than 90% of matter in the universe consists of this gas. A provocative question, however, is how the other elements originated.

Ever since nuclear physics began to develop in the 1920's and 1930's, it has been known that hydrogen can build up heavier atoms. Under certain conditions a number of hydrogen nuclei (protons) can combine into a heavier atomic nucleus (while emitting positrons to arrive at the proper charge). As was first shown by Bethe, similar reactions may take place in the interior of stars. The most common reaction is for four hydrogen atoms to fuse into a helium nucleus. The fusion generates tremendous energy, and it is by means of such fusion reactions that the stars get the energy which heats them and which radiates out into space.

Although the stars derive their energy from a buildup of helium, heavier elements are formed as "by-products" at the same time. The only condition that all these reactions require is heating of the hydrogen gas to a sufficiently high temperature, on the order of 10 million degrees or more. However, the temperature must not be too high: if it rises to several billion degrees, the heavier elements disintegrate and become hydrogen again. If the elements are to be built up fast enough, moreover, there must be just the right pressure in the area where the reaction is taking place. The perfect recipe for "cooking" elements from hydrogen gas, therefore, is suitable pressure plus suitable temperature.

It appears reasonable to assume that all the elements can be built up in the interiors of different kinds of stars; however, this does not solve the problem of how the elements originated, since they are also found in space. The matter of which the earth and other planets are composed cannot have been ejected from the sun. Rather, it is part of the matter of which the sun itself was originally made. When the sun condensed from a large cloud of plasma, most of the cloud was concentrated to form the sun. Smaller portions of the cloud were prevented by electromagnetic forces from falling into the sun; instead, they formed the planets, and they provide specimens of the original matter that went into the sun. Since the planets contain various amounts of the heavier elements, that ancient cloud must have contained them, too; they cannot have been cooked in the sun.

In principle the elements in that cloud may have been "cooked" before our solar system was born, in the interior of other stars antedating our sun. It has long been known that the stars eject gas into space. Our sun is continuously emitting plasma, and even more plasma emanates from novas and supernovas. It is doubtful, however, that such processes satisfactorily explain the existence of heavier elements in interstellar matter.

According to Gamow, the elements are "cooked" in his ylem during the *first hour* (sic!) after the moment of creation. When we allow for Gamow's own form of humor, we may assume that he was trying to illustrate how the elements could be formed under favorable conditions. If we apply his thesis to our model, a plausible conclusion would be that conditions are favorable for the formation of elements at a certain stage in metagalactic development. For example, temperatures and pressures for the synthesis of heavier elements could be attained in the formation of galaxies, in quasars, or in the separation of koinomatter and antimatter. But any such expectation from our model is no more than a guess.

We shall therefore conclude that the heavier elements may, in principle, be synthesized from hydrogen gas during metagalactic development. Whether this takes place in quasars, in the interior of stars, or in some other way is still uncertain. But since koinomatter and antimatter underwent the same development, the heavier koinoelements and antielements must have undergone the same process.

"CONTINUOUS CREATION"

We now give brief mention to the cosmological speculations of Bondi, Gold, and Hoyle, because they have received so much publicity. They start from what they call "the perfect cosmological principle," where the state of the universe has always been generally the same as it is now. The principle is already satisfied in part by the Charlier model, among others. But to fill in the gap they have introduced a new natural law, according to which neutrons—but not antineutrons—are created everywhere in space out of nothing. This process violates the law of conservation of energy. Further, their new natural law makes no provision for the symmetry between particles and antiparticles, which is fundamental to the physics of elementary particles, since it embraces the neutrons but leaves out the antineutrons. Attempts have been made to demonstrate the operation of the postulated process, but with negative results. When the theory of continuous creation is applied to metagalactic problems, it requires a mean density in the metagalaxy one hundred times greater than the probability yielded by observations. The distribution in space of the radio stars, and especially of the quasars, also conflicts with the theory. Its originators now seem to realize this and are trying to modify the theory.

However, as stated in Chapter I, the purpose of our book was to see how far we might go in our cosmological speculations without having to introduce new natural laws. In the light of this purpose, "continuous creation" falls outside the scope of this book.

SUMMARY AND CONCLUSIONS

We have considered one of the most important problems in natural science, yet also one of the most formidable. Its special character comes from our trying to gain insight into certain problems whose ultimate solution may not even lie within the scientific domain. Can we ever answer the question "How did the world begin?" in terms that do not beg more questions?

Here we are venturing on ground where emotional thinking rooted in philosophical and religious antecedents has traditionally dominated. We run a great risk of becoming bogged down in a jungle of myths, and it is the mixture of myths and science we must fear most of all. The lianas of myth suffocate the tree of science. In plain text, this means that if we try to overcome every barrier in our way by postulating a new natural law, then any line of reasoning is possible and any "explanation" is as good as the others. We must work to a fixed plan precisely because of the unsure and perilous ground under our feet.

Given these circumstances, it is reasonable to start from the natural laws discovered by physicists in their laboratories. We cannot be sure the laws apply throughout the cosmos, but it becomes a meaningful and well-defined task to analyze the cosmological problems within our stated terms of reference.

We have discussed three different models to describe the development of the universe—or, rather, of the metagalaxy.

1. According to the Lemaître-Gamow model, the primordial state is an "atome primitif" or "ylem," which explodes and ejects the galaxies in all directions. This model assumes that

    (a) the metagalaxy contains only koinomatter, no antimatter;

    (b) the universe was created at a specific moment in the form of a gigantic bomb.

On the other hand, an ylem state cannot be arrived at by previous contraction of the universe, as has often been suggested.

2. If we retain assumption 1(a), the present expansion of our metagalaxy could be derived from a previous contraction, but we cannot achieve the high concentration required by an ylem. This model assumes that
   (a) the metagalaxy contains koinomatter only;
   (b) the primordial state was a rarefied plasma (without antimatter), which contracted under the influence of gravitation.
3. Klein's model assumes that
   (a) there is a symmetry between koinomatter and antimatter;
   (b) the primordial state was a thin ambiplasma, contracting under the influence of gravitation.

If desired, this state can be led back to one even "more primordial," where there was only high-energy radiation (in turn giving rise to proton-antiproton pairs) in an otherwise empty space. A state of this kind might be regarded as the next simplest state of the universe. The simplest state, space without anything in it, is devoid of interest.

Starting with the primordial state, we can follow metagalactic development under the influence of gravitation up to the present state, though some of the processes—especially the separation of koinomatter and antimatter—are not yet satisfactorily analyzed. It seems possible that numerical agreement between the theory and observed values is attainable.

Klein's assumption of an originally homogeneous mixture of koinomatter and antimatter cannot be regarded as ad hoc, since the requirement of cosmological symmetry may be derived from the well-established findings of particle physics. Obviously, however, one is not an inevitable consequence of the other.

It would be possible to decide between the theories if we could demonstrate that antimatter does or does not exist in the universe, but this decision is beyond our powers at the present time. Several arguments may be marshaled for the existence of antimatter: radio stars, quasars, and other celestial objects with enormous energy outputs may be interpreted to indicate the presence of antimatter, but other interpretations are also possible. Furthermore, entire galaxies, say every second one in the universe, may consist of antimatter. It is even possible that every second star in our own galaxy may be an antistar.

When Einstein proposed the general theory of relativity, one of the conclusions he derived from it was that the universe could be finite and closed. He estimated its size at one billion light-years, an awesome figure when first put forth. However, that was prior to knowledge of metagalactic expansion, and we now know that the universe must be much larger. Moreover, the arguments for a finite and closed universe of metagalactic size have not proved very convincing. Calculations of the geometry of space pursuant to relativity theory are as easily reconciled with a space that is infinite and open. Enclosed space was an interesting possibility, but it does not seem to accord with reality. In any event, space must surely be much larger than the metagalaxy.

And there, in a sense, we have come back to the cosmological picture of the universe which governed before the relativity theories appeared, even though relativistic effects are of great importance in many respects. We can regard the metagalaxy as an event essentially taking place in quite ordinary three-dimensional space. The main features of Klein's theory can be described in three dimensions, even though the more detailed calculations require treatment in a four-dimensional curved space on strict mathematical grounds.

Acceptance of Klein's theory signifies that our picture of the universe covers a larger canvas. For Lemaître and Gamow, the metagalaxy is the whole universe. For Klein, our metagalaxy may possibly be only one of many others like it, and it is conceivable that these form an even larger system, which we have called a teragalaxy. Charlier's model of the universe thereby acquires renewed relevance. Assuming a primordial state of rarefied ambiplasma in a very large and perhaps infinite space, the ambiplasma can condense in many regions. One of these condensations developed into our own metagalaxy, but the same course of events could be enacted in other parts of space. Just as it dawned on us 50 years ago that our galaxy has many counterparts in space, we can now begin to speculate that our metagalaxy may not be a "loner" either. Our metagalactic sisters, however, would have to be very remote, on the order of trillions of light-years. And there we must stop, for we have already overstepped the bounds that we set for this study of the cosmological phenomena.

Reference: Essential parts of this book are taken from a scientific article by the author in Review of Modern Physics, **37**: *652(1965), entitled "Antimatter and the development of the metagalaxy."*

DR. WILLIAM K. WIDGER, JR.
RFD #2, Hillrise Lane
Meredith, N. H. 03253